MAR 31 2015

5

LONG SHOT

LONG SHOT

VACCINES FOR NATIONAL DEFENSE

KENDALL HOYT

HARVARD UNIVERSITY PRESS
Cambridge, Massachusetts
London, England *2012*

614.47
Hoy

Copyright © 2012 by the President and Fellows of Harvard College
All rights reserved
Printed in the United States of America

Library of Congress Cataloging-in-Publication Data

Hoyt, Kendall, 1971–
Long shot : vaccines for national defense / Kendall Hoyt.
p. ; cm.
Includes bibliographical references and index.
ISBN 978-0-674-06158-3 (alk. paper)
1. Vaccination—United States. 2. Vaccines—Government policy—United States. 3. Biological weapon—Safety measures—Government policy—United States. I. Title.
[DNLM: 1. History, 20th century—United States. 2. Vaccines—history—United States 3. Biological Warfare Agents—United States. 4. Security Measures—United States. QW 11 AA1]
RA638.H69 2011
614.4'7—dc23 2011026672

2

For Eli and Rhys

Contents

Text Acronyms

ADM	Advanced Development and Manufacturing
AEB	Army Epidemiology Board
AEI	American Enterprise Institute
AFEB	Armed Forces Epidemiology Board
AMS	Army Medical Graduate School
ARTP	Army Specialized Training Units
BARDA	Biomedical Advanced Research and Development Authority
BICIED	Board for the Investigation and Control of Influenza and other Epidemic Diseases
BIO	Biotechnology Industry Organization
BLA	Biologic License Application
BOB	Bureau of Biologics
BW	Biological Weapons
BWC	Biological Toxin and Weapons Convention
CDC	Centers for Disease Control
CBER	Center for Biologics Evaluation and Research
cGMP	Current Good Manufacturing Practices
CHPPM	Center for Health Promotion and Preventive Medicine
CMR	Committee on Medical Research
CWS	Chemical Warfare Service
DARPA	Defense Advanced Research Projects Agency
DHS	Department of Homeland Security
DMAT	Disaster Medical Assistance Team
DMS	Division of Medical Sciences
DOD	Department of Defense
DOE	Department of Energy
DTRA	Defense Threat Reduction Agency
EPA	Environmental Protection Agency
EPICON	Epidemiological Consultation

FDA	Food and Drug Administration
FOIA	Freedom of Information Act
FFRDC	Federally Funded Research and Development Center
GOCO	Government Owned Company Operated
GMP	Good Manufacturing Practices
HHS	Health and Human Services
IND	Investigational New Drug
IOM	Institute of Medicine
IP	Intellectual Property
JDO	Joint Development Office
JVAP	Joint Vaccine Acquisition Program
LSI	Lead System Integrator
MAP	Molecular Anatomy Program
MCM	Medical Countermeasure
NAS	National Academy of Sciences
NASA	National Aeronautics and Space Administration
NBSB	National Biodefense Science Board
NCI	National Cancer Institute
NDRC	National Defense Research Committee
NHSS	National Health Security Strategy
NIAID	National Institute for Allergy and Infectious Diseases
NIE	National Intelligence Estimate
NIH	National Institutes of Health
NRC	National Research Council
NSC	National Security Council
OSRD	Office of Scientific Research and Development
OSS	Office of Strategic Services
OTA	Office of Technology Assessment
PAHPA	Pandemic All-Hazards Preparedness Act
PhRMA	Pharmaceutical Research and Manufacturers Association
PHEMCE	Public Health Emergency Medical Countermeasures Enterprise
PHS	Public Health Service
PPP	Public Private Product Development Partnership
RFP	Request for Proposals
SAB	Science Advisory Board
SARS	Severe Acute Respiratory Syndrome
SGO	Surgeon General's Office
SNS	Strategic National Stockpile
UPMC	University of Pittsburgh Medical Center
USAMRIID	U.S. Army Medical Research Institute for Infectious Diseases
UNSCOM	United Nations Special Commission
VRC	Vaccine Research Center
VTEU	Vaccine Treatment and Evaluation Unit
WBC	War Bureau of Consultants

WHO	World Health Organization
WRAIR	Walter Reed Army Institute of Research
WRS	War Research Service

Archive Acronyms

AP	Aventis Pasteur Archives, Swiftwater, PA
MA	Merck Archives, Whitehouse Station, NJ
NA	National Archives, University of Maryland, College Park, MD
NAS	National Academy of Sciences, Committee on Biological Warfare Files, Washington, DC
LC	Library of Congress, Vannevar Bush Papers, Washington, DC
WR	Walter Reed Army Institute of Research, Joseph Smadel Reading Room Collection, Silver Spring, MD

LONG SHOT

Introduction

America faced a host of biological threats to health and security at the turn of the twenty-first century. Between 1990 and 2009, the United States contended with a foreign biological weapons program, bioterrorism, and a pandemic. Concerns about Saddam Hussein's biological weapon caches sent the U.S. military scrambling to immunize troops against smallpox and anthrax. The 2001 anthrax attacks demonstrated that non-state actors could terrorize civilian populations with biological weapons as well. A strange outbreak of severe acute respiratory syndrome (SARS) in 2002 and new avian and swine flu strains provided further reminders of the pandemic potential of infectious diseases.

While each individual threat posed limited danger to the nation, together they raised awareness of the catastrophic potential of disease and spurred large-scale government demand for vaccines to defend soldiers and civilians. Even so, only one new biodefense vaccine was licensed during this period. Many candidates were technologically feasible and well funded. A next generation anthrax vaccine, for example, has been a top priority for the U.S. government since the first Gulf War, but twenty years and over a billion dollars later, the United States still does not have this new vaccine.

These vaccine development failures are startling. The United States was once capable of mounting rapid development campaigns in response to national emergencies. World War II–era programs generated ten new or improved vaccines against diseases of military significance. In some cases, these programs produced new vaccines in time to meet the objectives of specific military operations. Botulinum toxoid, for example, was mass-produced before D-day in response to (faulty) intelligence that Germany had loaded V-1 bombs with the toxin, and a Japanese encephalitis vaccine was developed in anticipation of an Allied land invasion of Japan.

The ability to develop new vaccines quickly is essential to national security and public health. As biological threats proliferate and new diseases continue to emerge, we have an urgent need to understand the conditions that foster timely innovation. Present-day vaccine programs tinker with push and pull policies (such as research grants or market guarantees) to spur innovation, but they rarely scrutinize the development process itself to address critical obstacles to innovation. *Long Shot* examines the developmental history of vaccines to uncover the conditions that first drove, and later inhibited, vaccine innovation.

This historical investigation provokes important questions for today: What factors foster timely vaccine innovation? What are the dynamics of industrial decision making during national security crises? How can we generate innovation for medicines that are socially valuable but commercially unappealing? How have military-industrial partnerships changed, and why does this matter for vaccine development? And finally, how can history inform efforts to rebuild biodefense capabilities in the twenty-first century?

~ SECURITY EXPERTS OFTEN refer to disease as a "nontraditional" threat, but few security threats are more traditional than disease. For the military, fighting disease has always been an important corollary to fighting the enemy. History is rife with battles in which bugs played a larger role than bullets.[1] Whether at peace or at war, military settings encourage high rates of disease. Military camps breed new diseases and magnify the effects of common diseases. Armed conflict exacerbates these conditions, producing social dislocations and generating populations of wounded and stressed individuals that boost the incidence and spread of disease.[2]

Intentional disease threats are equally "traditional." Armies have employed rudimentary forms of biological warfare since antiquity, when retreating soldiers would contaminate enemy wells with human and animal corpses.[3] In the Middle Ages, Tartar invaders gained notoriety for catapulting plague-infected corpses over city walls.[4] By the mid-twentieth century, delivery methods had improved considerably as the United States, the Soviet Union, the United Kingdom, France, and Japan all invested in biological weapons programs. The United States suspects that countries such as China, Russia, Iran, Syria, Cuba, and North Korea have made more recent investments.[5]

Several cults (the Rajneeshees and Aum Shinrikyo) and terrorist groups (al-Qaeda) have demonstrated an interest in biological weapons as well. While these non-state actors have had limited success, past performance is not predictive of future outcomes. Biological weapons will remain inherently attractive to dissatisfied groups seeking asymmetric advantages because these weapons are less expensive and more difficult to trace than nuclear weapons. Over time, supply will grow to meet demand as the economic, technical, and educational obstacles continue to erode.

Opportunities for natural diseases to emerge and spread are increasing as well. Population growth, climate change, and expanding travel and trade patterns create new opportunities for disease. The widespread use and misuse of antibiotics and antivirals also play a role in breeding new generations of pathogens that evade medical arsenals. Approximately eighty new diseases have emerged or reemerged since 1970 in response to these evolutionary pressures.

While disease threats increased toward the end of the twentieth century, vaccine innovation rates declined. Few understand the seriousness of this problem because widespread errors in official vaccine license records create the false impression that innovation has steadily increased over time. These data create support for industrial innovation policies that have not worked well for many vaccines.

To develop a more accurate picture of innovation patterns, I restored original licenses that had been lost from federal records, corrected inaccuracies, and identified licenses issued for noninnovative activity. These new data demonstrate that innovation has been falling, not rising, since World War II. These data also demonstrate that the historical record is inconsistent with prevailing theories of innovation in industrial settings.

Historians often argue that technological innovation is a function of economic incentives, individual firm capabilities, and the available stock of "scientific knowledge" or "technological opportunities."[6] Recent investigations of vaccine development also interpret innovation as a function of commercial developers responding to technological opportunities to maximize profit.[7] My research reveals that the three key factors in market-based theories of innovation are at odds with the data: economic incentives, firm capabilities, and the stock of technological knowledge/opportunities were weaker for the vaccine industry in the 1940s and 1950s, when innovation rates were high, and stronger in the 1980s and 1990s, when innovation rates were low.

Vaccines do not always lend themselves to standard interpretations, in part because markets often fail to inspire socially optimal levels of vaccine innovation and consumption.[8] This is particularly true for global health vaccines (sold to developing countries that cannot afford to pay premium prices) and biodefense vaccines (used on a limited basis or stockpiled). Demand for these vaccines is often insufficient to ensure socially desirable levels of innovation and supply.

Given the inherent disincentives for developing biodefense vaccines, health and defense planners should ask themselves not "Why are innovation rates falling?" but "Why did the system ever work at all?" and, more specifically, "What factors permitted an effective industrial response to the government demand for vaccines in the 1940s and 1950s, when theoretical studies tell us that innovation was relatively less likely than it is today?"

Wartime development programs were not a triumph of scientific genius, but of organizational purpose and efficiency. Often, the scientific foundation for a vaccine had been established years, if not decades, in advance of its development. It was not until World War II, however, that many of these concepts were plucked from the laboratory and developed into working vaccines. Wartime development programs excelled at consolidating and applying preexisting knowledge for the purpose of product development. To accomplish this task, innovators employed an integrated approach to research and development.

Integrated is my term for research that is managed from the top down, integrated across disciplines and developmental phases, and situated in a community that facilitates information exchange and technology transfer. Integrated research is similar to, but distinct from, other commonly used terms such as "champion-led" and "translational" research. Integrated research resembles champion-led research in that project directors coordinate development teams from the top down and look beyond traditional job descriptions, organizational routines, and funding sources to overcome obstacles that arise in product development. In both cases, project directors must have the resources and authority to coordinate activities along developmental phases, the skills to manage projects across disciplines, and the opportunity to collaborate with early developers and current users to incorporate their insights. Unlike champion-led research, however, integrated research is not devoted to one product or approach; project directors continuously review and reassess a portfolio of alternative approaches.

Integrated research is also similar to translational research in that it applies basic research insights to practical development problems. Unlike some definitions of translational research, integrated research is not exclusively concerned with moving laboratory research into clinical trials. Practical applications can take the form of new research agendas, methods, protocols, products, and processes.

Integrated research goes beyond both definitions to describe a specific governance structure, research method, and cultural context for research.

Top-down governance facilitates vaccine development because it allows developers to coordinate activities across a wide range of disciplines while maintaining focus on long-term development goals. When project directors manage a vaccine across disciplines and development phases, they also maintain strong situational awareness of product development needs that permits accurate and timely decision making. This approach differs from NIH- and academia-supported investigator-initiated research, which seeds discovery from the bottom up. Bottom-up processes are essential in the discovery phase, but they can be counterproductive in the later stages of development.

Integrated research does not only govern from the top down; it defines problems and combines work processes across disciplines and developmental phases. Epidemiologists, clinicians, lab scientists, bioprocess engineers, regulators, and manufacturers work closely with one another at different stages to understand the upstream and downstream requirements of their collaborators. Integrating work streams in this fashion introduces unforeseen efficiencies and insights.

Integrated research thrives in the context of informal collaborative communities that facilitate the transfer of tacit knowledge. Just as informal guilds allow artisans to craft products with greater efficiency and skill, research communities with strong working relationships can translate results and negotiate solutions more effectively together than they could in isolation. Cognitive anthropologists have observed that these informal networks—or "communities of practice"—can facilitate knowledge creation and transfer.[9] Communities of practice engage in activities such as collective problem solving, information and asset sharing, on-site collaboration, and knowledge mapping.[10]

This cultural milieu is particularly important for vaccine development because the knowledge required to grow and manipulate biological material is often context-dependent (i.e., the techniques that allow a pathogen to grow well in one lab, may not work in another lab, or at a different volume). This type of knowledge is what R&D management experts call "sticky." Information is sticky when it "is costly to acquire, transfer, and use in a new location."[11] Sticky information is most easily transferred among individuals with a high degree of trust and familiarity.

While integrated research practices have antecedents in prewar industrial settings like Bell Labs and General Electric, these methods gained greater currency during World War II when the military, industry, and academia joined forces to mobilize research and development activities.[12] Vaccine development programs were most successful when they drew on the military's direct experience with a particular disease. Just as military interest in weapons manufacturing, information systems, and machine control inspired innovation, so, too, has military interest in the problems of disease control.[13] Tremendous advancements in public health and vaccine development came from a longstanding military preoccupation with pathogenic threats. The U.S. military, in particular, made significant contributions to over half of the vaccines developed in the twentieth century.[14]

The success of these ventures can be attributed in part to the military's status as a "lead user" of vaccines. Lead users are "organizations or individuals that are ahead of market trends and have needs that go far beyond those of an average user."[15] Pairing lead users with developers encourages innovation because these teams yield unique insights and rapid solutions. World War II development programs paired lead users (the military) with developers (academia and industry), further integrating research and development. It should not be surprising, therefore, that this collaboration yielded a significant number of new or improved vaccines. The surprise is that the military was able to engage industry in these projects. *Long Shot* examines historical efforts to develop vaccines for national defense to reveal the factors that drove innovation when financial returns were low but social returns were high.

LONG SHOT IS DIVIDED into seven chapters. In chapter 1, I chart the rise of disease threats and discuss why certain pathogens threaten national security. I outline the strategic value of vaccines relative to other available measures to deter and defend against disease threats. While a robust defense hinges on many factors, vaccines continue to play an essential role in defense plans. I propose a strategy that will allow vaccine development capabilities to keep pace with evolving threats.

Chapter 2 outlines the theoretical framework of the book. Previous studies of vaccine innovation have had to contend with inaccurate vaccine license data. I introduce a more comprehensive, historically accurate data set and investigate its implications for prevailing theories on the sources of innovation in the vaccine industry. I argue that market-driven theories of innovation are insufficient to explain the historical patterns observed, and suggest a larger role for nonmarket factors.

Chapter 3 examines World War II vaccine development programs to reveal how the military, in collaboration with academia and industry, achieved unprecedented levels of innovation to counter war-enhanced disease threats. While the war focused funding and attention on the immediate problem of getting new vaccines to the troops, the nature of the collaboration (among scientists, administrators, and industrialists) mattered as much as, if not more than, the level of urgency surrounding the mission. Wartime development programs forged new research partnerships and practices that generated a record number of new vaccines.

High rates of innovation persisted in the postwar era, even after the urgency and structure of wartime programs were gone. Chapter 4 demonstrates that participation in wartime programs forged a set of personal friendships, institutional affiliations, and research practices that sustained innovation. Formal military-industrial research and development partnerships gave way to informal collaborative networks that influenced everything from which vaccines were developed to how to develop them. An enduring sense of patriotism, social obligation, and familiarity supported these networks and influenced industry investments. A close examination of the collaborative relationships between the Walter Reed Army Institute of Research and commercial vaccine manufacturers (Merck & Company and the National Drug Company in particular) illustrates how the military continued to influence innovation during this period, even when the economic logic for developing a particular vaccine was not compelling.

Chapter 5 illustrates the legal, economic, and political transformations that disrupted military-industrial networks in the 1970s and 1980s. After the Vietnam War, military research organizations restructured and the vaccine industry consolidated. Vaccine development activities migrated to the National Institute of Health and to academia where publications—not products—were prized. Rapid advances in the biosciences fractured the discipline into multiple subspecialties and scattered expertise across a wider range of research institutions and smaller biotechnology companies.

Ironically, many late-twentieth-century developments that have been celebrated as a boon for innovation—the explosion of scientific subdisciplines and bioengineering techniques, the specialization and dispersion of firm capabilities, and the growth of outsourcing—have frustrated efforts to employ integrated research practices. The present environment gives developers access to a wider range of scientific expertise and sophisticated techniques, lower overhead, risk-sharing arrangements, and near-term market efficiencies, but it also leads to higher overall development times and failure rates. Integrated research practices proved harder to pursue in this environment and innovation rates began to fall.

The events surrounding 9/11 mobilized the federal government and the pharmaceutical industry with a renewed sense of urgency and a spirit of cooperation reminiscent of World War II. However, chapter 6 demonstrates that urgency alone was insufficient to spur innovation without the focus of integrated research programs. Federal programs spent billions of dollars on vaccine development but they failed to support the nonmarket factors that mattered most. Instead, federal biodefense investments reinforced the balkanization of vaccine research and development that has suppressed innovation for the last several decades.

The landscape for vaccine development has changed irrevocably since the 1970s, and it is neither possible nor desirable to reproduce midcentury formulas for success. History does, however, offer important lessons for our efforts to rebuild medical countermeasure development capabilities. Chapter 7 proposes a new direction for biodefense research and development that builds on insights from historically successful vaccine development programs.

1 Disease, Security, and Vaccines

WELL BEFORE THE anthrax letter attacks in 2001, a series of events in the 1990s alerted security experts to the growing risk of biological attacks on U.S. soil. In 1992, Ken Alibek, a bioweapons scientist from the former Soviet Union, defected to the United States and provided hair-raising accounts of biological weapons (BW) development in his country. Working under the cover of a government pharmaceutical agency (Biopreparat), the Soviet Union had operated an offensive biological weapons program since 1972, the same year it signed a treaty banning the development of these weapons.[1]

Russian President Boris Yeltsin banned this program in 1992, forcing large numbers of bioweapons scientists to seek new employment. A 1997 visit to one of Russia's largest bioweapons labs (Vektor in Koltsovo, Novosibirsk) revealed that what was once a high-security compound containing thirty buildings and nearly four thousand employees had been reduced to "a half-empty facility protected by a handful of guards who had not been paid for months."[2] This facility still contains what is supposed to be the only cache of the smallpox virus outside of the U.S. Centers for Disease Control and Prevention (CDC). This visit led arms control experts to fear that samples of the virus might have escaped to other countries, along with some well-trained bioweapons scientists, making it easier for other state and non-state actors to obtain biological weapons.

By 1995, the United Nations Special Commission (UNSCOM) confirmed that Iraq had produced and filled anthrax and botulinum toxin in bombs, rockets, and airplane spray tanks for the first Gulf War (1990–1991). That same year, the Clinton administration learned that non-state actors also had an interest in mass casualty weapons and methods when Aum Shinrikyo attacked a Tokyo subway with sarin gas in March, and domestic terrorists bombed a federal building in Oklahoma City in April.[3] Investigations into the activities of Aum Shinrikyo revealed that the cult had also attempted multiple biological attacks between 1990 and 1995, including attacks on U.S. naval bases, Narita Airport, the Imperial Palace, and the Japanese Diet.[4]

Aum was not the first cult to show interest in BW. In another well-known case, the Rajneeshee cult contaminated salad bars in Oregon with *Salmonella typhimurium* to influence turnout for a local election in 1984, sickening nearly a thousand people. When state and federal agents investigated the cult's commune on unrelated charges a year later, they uncovered additional plots and found germ bank invoices for a variety of other pathogens, including *Salmonella typhi* and *Francisella tularensis.*[5]

Empirical studies from the 1990s reveal a convergence of two unsettling trends: (1) the proliferation of bioweapons materials and expertise and (2) the increased propensity of terrorist groups to inflict indiscriminate mass casualties. The Chemical and Biological Weapons Nonproliferation Project at the Monterey Institute's Center for Nonproliferation Studies maintains a database of publicly known attempts to acquire or use chemical, biological, radiological, or nuclear materials.[6] The database shows that terrorist incidents have been on the rise since 1985, with peaks for the use of chemical and biological agents in 1995 and 1998. Preferred targets have changed with increased emphasis on civilian populations, and biologic agents tend to be associated with nationalist or separatist objectives, the desire to retaliate, exact revenge, or to fulfill apocalyptic prophecies.

These findings highlight the emergence of a new breed of terrorist prone to indiscriminate violence. In one characterization, this new type of terrorist is "less interested in promoting a political cause and more focused on the retribution or eradication of what he defines as evil . . . For such people, weapons of mass destruction, if available, are a more efficient means to their ends."[7] Increasingly, this breed populates religious groups, racist and antigovernment groups, and millenarian cults.

Databases that track the composition and activities of international terrorists since 1968 reveal a growing number of religiously motivated groups and a link between religious motivations and higher levels of violence. For example, religious groups were responsible for only 25 percent of the terrorist acts committed in 1995, and yet they accounted for 58 percent of the fatalities.[8] One member of Hezbollah provided a succinct explanation for this phenomenon: "We are not fighting so that the enemy recognizes us and offers us something. We are fighting to wipe out the enemy."[9]

Naturally occurring disease threats are growing as well. After decades of watching disease rates fall, experts note that the death rate from infectious disease has been on the rise since 1980. Even if one excludes HIV/AIDS cases, the death rate from infectious disease rose by 22 percent in the United States between 1980 and 1992.[10] A 1992 Institute of Medicine (IOM) report identified eight factors leading to the introduction and spread of infectious disease and concluded that each factor was trending upward.[11] These factors include population growth, globalization (immigration, travel, and commerce), changing land-use patterns, climate change, and changing health-care practices, such as the use of immunosuppressive therapies, more invasive medical procedures, and the overuse of antibiotics.

By 2000, the National Intelligence Estimate (NIE) revealed that the overall number of U.S. infectious disease deaths had doubled since the 1980s.[12] The NIE report concluded that this trend would continue: "As a major hub of global travel, immigration, and commerce with wide-ranging interests and a large civilian and military presence overseas, the United States and its equities abroad will remain at risk from infectious diseases."[13]

In addition to newly emerging infectious diseases, over twenty known diseases (such as tuberculosis, malaria, and cholera) have spread to new geographic areas since 1973.[14] In most cases, these diseases have reemerged in more virulent and drug-resistant forms.

For example, the United States wiped out all cases of domestic malaria in the 1960s, but high rates of immigration and international travel reintroduced the disease, with approximately 1,300 new cases per year.[15] Without an effective vaccine, and in the face of growing drug resistance, the geographic spread of malaria is increasingly difficult to control. According to one estimate, "the first-line drug treatments for malaria are no longer effective in over 80 of the 92 countries where the disease is still a major health problem."[16]

AS THE EVIDENCE OF disease threats mounted, a handful of health experts and security planners began to explore the links between global health issues and U.S. national security interests. Clinton's national security advisor, Samuel Berger, explained: "A problem that kills huge numbers, crosses borders, and threatens to destabilize whole regions is the very definition of a national security threat."[17] In 1998, President Clinton appointed Kenneth Bernard (former U.N. international health attaché) to a newly created senior post on the National Security Council (NSC) to examine disease threats to security. This move, according to Bernard, was evidence that "We have broken out of the standard approach of dealing with international health issues, which have been seen as solely health or humanitarian issues. Now it is seen in the broader context of national security issues and economic development."[18]

The concept of disease as a national security threat received ample support in the 2000 NIE report, which grappled with the impact of global infectious disease threats on U.S. interests.[19] In that same year, the NSC and the U.N. Security Council formally classified HIV/AIDS as a national security threat, placing it on a list alongside weapons of mass destruction. Rising rates of infection in Africa, South Asia, and Russia, they reasoned, contributed to humanitarian emergencies and military conflicts that would inhibit democratic development abroad. In 2006, as highly pathogenic H5N1 flu strains circulated among bird populations, the Bush administration also placed pandemic disease on a list of national security threats and appropriated billions to prepare for a pandemic.[20]

The Obama administration formally recognized the connection between health and security as well through the National Health

Security Strategy (NHSS), drafted by the Department of Health and Human Services (HHS) in December 2009.[21] The goal of the NHSS is to strengthen the nation's ability to prevent, respond to, and recover from large-scale health threats arising from bioterrorism, pandemics, or natural disasters. The NHSS does not make the case for disease as a security threat in and of itself, but simply states: "The health of the nation's people has a direct impact on that nation's security."[22]

The political advantages of defining health initiatives in terms of national security are clear. Defined as such, select HHS projects gain higher priority and a stronger claim on limited resources.[23] The basic claim, however, deserves more attention: *Does disease pose a national security threat?* Not everything that affects the health of the nation's people has a direct impact on security. Far more people die of traffic accidents than influenza or pneumonia in any given year, but no one believes that auto safety is a matter for the Department of Defense.[24] A working definition of biosecurity threats should distinguish between diseases and events that threaten not just the individual but the state as a whole.

Richard Danzig, former secretary of the navy, discriminates between contained and uncontained catastrophes. He argues that—unlike bombings or hurricanes, which represent "contained catastrophes"—natural or manmade plagues can destabilize our way of life and "call into question our near-term abilities to recover."[25] Broadly speaking, disease threatens national security when it impairs the ability of a state to maintain order, defend itself, and project power. Defined in this way, few diseases rise to the level of a national security threat. The event (i.e., a BW attack, disruption of the food supply through contamination or disease, a human epidemic) would have to be both widespread and severe to disrupt essential state functions.

Not all biological threats would have to meet this catastrophic standard to classify as a security threat, however. Political scientist Gregory Koblentz provides a taxonomy of state-level threats that places pandemic diseases alongside biological terrorism and dual-use research areas (e.g., genetic engineering and synthetic biology).[26] Within this rubric, intent matters as much as outcome. The U.S.

government considers most forms of terrorism a security threat regardless of the means (guns, bombs, or germs) or the outcome of a particular plot. Unlike crime, which is committed for private gain, terrorism seeks political, economic, or religious change.[27] A pandemic may harm a larger number of U.S. citizens than a single act of bioterrorism, but because terrorists intend to threaten U.S. policies and institutions, their activities cross the national security threshold, even if they are relatively unsophisticated.

War, disease, and state power are strongly associated, but there are few systematic investigations of the causal relationships among these forces.[28] Andrew Price-Smith tackles this subject through a series of case studies in which he demonstrates that disease can have a direct influence on military operations and state power.[29] High disease rates, he argues, can destabilize the state, raising the risk of intrastate war. Widespread disease can be both the result of and the cause of a state's failure to provide essential goods and services. This can lead to fear, violence, and rioting against the state. Both directly and indirectly, therefore, disease erodes the perceived legitimacy of the state. In Zimbabwe, for example, where over 30 percent of the population is expected to die from HIV/AIDS between 2005 and 2015, the virus has weakened military institutions, law enforcement, economic growth, and productivity. In this manner, disease affects the ability of Zimbabwe to maintain order, defend itself, and project power.[30]

Disease can also amplify the impact of war and in some cases influence the outcome of military conflicts. Price-Smith uncovers mortality data from German and Austrian archives to demonstrate that the 1918 influenza pandemic affected combatants unequally and influenced the outcome of World War I. War, he hypothesizes, facilitated the spread and augmented the lethality of the virus. The pandemic, in turn, hastened the end of hostilities by undermining German offensive operations in the spring and summer of 1918.[31] The flu hit the Central powers first and crippled military operations by elevating mortality and morbidity, reducing morale, raising costs, and limiting the flow of troops and material to the front.

~ IF ONE ACCEPTS the premise that disease on a certain scale can pose a security threat, it then becomes clear that effective methods of disease control confer strategic security advantages. Vaccines offer one of the oldest and most effective methods of disease control. Military organizations have exploited this means of protection since rudimentary methods of vaccination became available in the 1700s. General Washington famously variolated the entire Continental army in 1777 to protect against the ravages of smallpox during the Revolutionary War.[32] A smallpox outbreak from the year before had been a leading factor in the failure of the Continental army to capture Quebec.[33] While variolation was a highly controversial procedure with severe adverse effects (including death), Washington determined that the promise of immunity conferred strategic advantages that outweighed the risk of the procedure.[34]

Over two centuries later, vaccines continue to play an important role in defense, diplomacy, and deterrence. And yet the national security community often overlooks the importance of vaccines and/or takes their availability for granted. This is due partly to a relatively weak political constituency for biodefense issues (Chapter 7) and partly to the nature of vaccines themselves. Unlike a new weapons system that awes onlookers with a display of fireworks, when vaccines work, nothing happens.

U.S. strategy to reduce biological weapons vulnerabilities has consisted of three tactics: nonproliferation, counterproliferation, and defense—or "consequence management."[35] This strategy recognizes that no single tactic is sufficient to address the threat, and all must be pursued simultaneously. Although all three are useful, weaknesses in preventive tactics necessitate a stronger emphasis on defense.

U.S. BW nonproliferation policy rests on a collection of arms control measures that have failed to prevent the spread of biological weapons technology due to the twin challenges of monitoring and enforcement. The Biological Weapons Convention (BWC) lies at the heart of U.S. nonproliferation efforts. The BWC is an international treaty (with 161 state parties and fourteen signatory states) that outlaws biological arms. In 2001, the Bush administration withdrew support from efforts to strengthen the verification protocol, arguing that the draft put forth at the Fifth Review Conference would not effectively limit proliferation. The Obama administration upheld this decision at an annual review conference in 2009. Given the inherent challenge of devising an enforceable verification protocol, legally binding multilateral treaty solutions in this area remain feeble.[36]

Apart from the BWC, the U.S. government supports export controls through the Australia Group, a committee composed of forty countries and the European Commission. Through a nonbinding agreement, participating countries limit the export of materials and technologies relevant to the production of chemical and biological weapons to proliferant countries.[37] However, given the dual-use nature of the materials controlled under this agreement and the global expansion of the biotechnology and pharmaceutical sectors, export controls of this nature are unlikely to be effective over the long term.[38]

In the future, nonproliferation policy could encompass new international laws. The Harvard Sussex Program on Chemical and Biological Weapons Armament and Arms Limitation has called for an international convention to criminalize chemical and biological weapons development and related activities.[39] Such a law would promulgate a valuable normative prohibition against biological weapons and facilitate international cooperation for nonproliferation. However, it may be difficult for international courts to collect sufficient evidence to prosecute suspected violators. Historical intelligence failures (the United States seriously underestimated Soviet BW capabilities during the Cold War and overestimated Iraq's prior to the second Gulf War) reveal that it can be exceedingly difficult to verify biological weapons activity.[40]

U.S. BW counterproliferation policy relies on deterrents, ranging from surveillance and interdiction, political persuasion, and the threat of military force to preempt or respond to a biological attack.[41] All of these strategies have limited utility because biological weapons easily escape early detection, favor the attacker, and are difficult to trace.[42] These deterrents are particularly weak for nonstate actors, who are less sensitive to state-level interventions such as political persuasion and military action.

Given our inability to limit the spread of bioweapons-relevant capabilities, and the inexorable evolution of pathogens, security planners have a clear need to emphasize defense. Phillip K. Russell, former commander of the U.S. Army Medical Research and Development Command, argues that the success of any response to pathogenic threats will depend on "the rapidity of the public health response, the effectiveness of a vaccination campaign, and, most importantly, the availability of vaccine."[43]

Many have questioned this relatively vaccine-centric view of biodefense, particularly as it applies to civilian preparedness. Vaccines are just one piece of a multipronged response that requires a system to detect and diagnose disease and to distribute and administer medicine. Vaccines are most effective when they are given in advance so that individuals have time to build immunity. Military personnel and first responders are able to have advance warning in some cases, but vaccines offer an impractical first line of defense for civilians.

These arguments are valid, but some critics take them a step further contending that the United States should shift its focus away from vaccines altogether to emphasize postexposure prophylactics (such as antibiotics), physical barriers (such as face masks), and social defenses (such as quarantine). While these measures are an important part of biodefense plans, this position overlooks the fact that vaccines offer a vital second line of defense in postattack scenarios. It may be impossible to predict and preempt an attack, but vaccines can mitigate the scope and duration of an event, even if they are administered *post hoc*.

Vaccines have strategic advantages in many *post hoc* scenarios. First, they allow other defensive measures to work more effectively. Efforts to build surge capacity in the health-care system, devise quarantine protocols and decontamination procedures, and employ protective equipment, like building filters and respiratory gear, will all have greater success if health-care workers, emergency responders, and civilians have access to effective, fast-acting vaccines. For example, vaccinated responders can enter biologically contaminated areas to triage victims and administer vaccines, which will shorten quarantines. Second, vaccines can also be used to protect surrounding populations from secondary waves of infection. The 2009 H1N1 pandemic offers a good example of how this works. The United States was susceptible to the first wave of infection in the spring, but vaccines were developed by late fall in time to protect high-risk populations against the second wave of infection when flu season returned. Third, some vaccines, such as the smallpox vaccine and the anthrax vaccine (in combination with antibiotics), can be used for postexposure prophylaxis. Fourth, vaccines would be valuable in reload scenarios where terrorists might attempt to threaten civilian populations with repeated, undetected attacks.[44] As with pandemic flu, health responders would miss the first wave of attacks, but they could, in theory, vaccinate against subsequent attacks that used the same pathogen.

Therapeutic countermeasures—such as antivirals and antibiotics—are more attractive to biodefense planners in many ways. They can defend against a wider range of pathogens and address numerous natural disease threats, they simplify stockpile requirements, and they are useful in a range of postattack scenarios. These broader-spectrum countermeasures may also offer the only recourse if populations are attacked with multiple agents simultaneously.

There are, however, no magic bullets in biodefense. While broad-spectrum antibiotics and antivirals should be an important development goal in any biodefense program, the widespread use of these drugs may breed resistance. When they are used liberally across an entire population, as would be the case in a large-scale emergency (or even in a moderate flu pandemic), they could breed resistance more quickly than normal use patterns would predict.

New antibiotics and antiviral therapies may also be more difficult to develop in the near term than new vaccines.[45] Limited industry interest has weakened the pipeline for new antibacterial drugs. Furthermore, the usefulness of these drugs is limited to bacterial pathogens, which account for only 29 percent of the HHS's select agent list.[46] Viral threats dominate, but antivirals are few and far between. Currently only one antiviral—Cidofovir—is indicated for smallpox, and this drug is of limited use in a large-scale emergency. It may be ineffective in symptomatic patients, it must be administered intravenously, and it costs $2,000 to $5,000 per person. A smallpox vaccine, on the other hand, is generally easier to develop, easier to administer in an emergency, and costs $30 to $40 per person.[47]

In addition to defensive value, vaccines have diplomatic value. Japanese encephalitis and polio vaccines, for example, played an important role in the battle for hearts and minds during the Cold War.[48] The U.S. military was able to gain trust and curry favor by offering these vaccines to populations in contested territories. If the United States could donate an HIV, malaria, tuberculosis, or pandemic flu vaccine today, this gesture would yield significant diplomatic benefits and political leverage in international negotiations. Conversely, should the United States fail to develop these vaccines while another country succeeds, the United States might be forced to yield financial and strategic assets in exchange for access to a vaccine.[49]

Vaccines have additional strategic value as a deterrent.[50] Simply having a smallpox vaccine in the Strategic National Stockpile (SNS), for example, reduces the potential effectiveness of that virus in the eyes of the attacker and thereby reduces its attractiveness as a weapon. With the addition of each new vaccine to the SNS, an attacker is forced to search for new pathogens that are likely to be more difficult for them to obtain, handle, or weaponize.

While vaccines are critical strategic assets, health and security planners cannot assume that they will be available when they are needed. In the immediate aftermath of the 2001 anthrax attacks, Michael Friedman, a former FDA administrator enlisted by the Office of Homeland Security to coordinate industry biodefense efforts, took comfort in the muscle of the pharmaceutical industry. He boasted, "A lot of people would say we won World War II with the help of a mighty industrial base. In this new war against bioterrorism, the mighty industrial power is the pharmaceutical industry."[51] In reality, however, vaccine development capabilities have eroded steadily since the 1970s, rendering innovation less likely and supplies more vulnerable.

Vaccine innovation rates have not merely fallen (Figure 2.1, Chapter 2), they have failed to keep pace with the growing number of disease threats facing the United States (Tables 1.1 and 1.2). New vaccines are available for only three out of the fifty newly emerging disease threats identified since 1973. Since the 1970s, the FDA has

Table 1.1 Development Status of Vaccines for Emerging Diseases

Appearance	Disease	Vaccine
1973	Rotavirus	Two vaccines licensed in 2006 ar 2008. (A 1998 vaccine was recall for causing bowel obstruction in infants).
1974	MRSA (Methicillin Resistant Staphylococcus)	
1975	Parvovirus B19	
1976	*Vibrio vulnificus*	
1976	*Cryptosporidium parvum*	
1977	*Clostridium difficile*	
1977	Hantaan virus	
1977	Delta viral hepatitis	
1977	*Campylobacter* sp.	
1977	Ebola virus	
1977	*Legionella pneumophila*	
1979	*Cyclopsora cayetanesis*	
1980	Human T-lymphotrophic virus I (HTLV-I) t-cell lymphoma/leukemia	
1981	*Staphylococcus aureus*—toxin producing (TSS)	
1982	*Escherichia coli* O157:H7	
1982	HTLV-II	
1982	*Borrelia burgdorferi* (Lyme disease)	Vaccine licensed in 1998. Pulled the market in 2002.
1983	HIV	
1983	*Helicobacter pylori*	
1984	*Hemophilus influenzae aegyptius*	
1985	*Enterocytozoon bieneusi*	
1985	Vancomycin Resistant Enterococcus (VRE)	
1986	*Chlamydia pneumoniae*	
1988	Human herpesvirus 9	

89	*Rickettsia japonica*	
89	Hepatitis C	
90	*Balamuthia madrillaris*	
90	Human herpesvirus 7	
90	Hepatitis E virus	
92	*Vibrio cholerae O139* (new strain associated with epidemic cholera)	
92	*Bartonella henselae*	
92	*Rickettsia honei*	
92	*Tropheryma whippelii*	
92	Barma Forest virus	
93	Sin Nombre hantavirus	
94	*Anaplasma phagocylophilum*	
94	Hendra virus	
96	Human herpesvirus 8	
96	variant Creutzfeldt-Jakob disease	
97	VISA/VRSA (Vancomycin-Intermediate Resistant Staphylococcus)	
97	*Rickettsia slovaca*	
99	*Cryptococcus gattii*	
99	*Ehrlichia ewingii*	
99	Nipah virus	
01	Human metapneumovirus	
02	SARS-CoV	
03	H5N1 avian flu	Vaccine licensed in 2007
04	Monkeypox virus	
08	Lujo virus	
09	H1N1 swine flu	Vaccine licensed in 2009

Note: This list does not include reemerging disease threats, such as malaria and tuberculosis.
Sources: "Accelerated Development of Vaccines," *Jordan Report* (2000), www.niaid.gov/publications/
/jordan.pdf; World Health Organization, *World Health Report* (Geneva: 1999); David F. Gordon, Don
ah, and George Fidas, "The Global Infectious Disease Threat and Its Implications for the United
tes," *National Intelligence Estimate* 99–17D (January 2000); D. Morens, G. Folkers, and A. Fauci,
nerging Infections," *Lancet Infectious Disease* 8 (2008): 710–719; A. Barrett and L. Stanberry, eds., *Vac-*
s for Biodefense and Emerging and Neglected Diseases (Academic Press, 2009), 6–7.

Table 1.2 Late-Stage Development Status of Vaccines for Category A Biological Threats

Disease	Vaccine	License Status	Last Year Produced
Anthrax	Emergent BioSolutions (Anthrax Vaccine Adsorbed)	BioThrax: Relicensed the 1970 version in 2002	Current
	Emergent BioSolutions (rPA)	Phase II	
	PharmAthene (rPA)	Phase II (completed)	
Botulism	DOD pentavalent toxoid for serotypes A-E	IND (filed 1979)	1995
	DynPort (recombinant A/B)	Phase II	
Plague	Greer inactivated vaccine	Licensed, ceased manufacture	1998
	DynPort (Recombinant subunit)	Phase II	
Smallpox	Wyeth calf lymph vaccinia vaccine	Dryvax: Licensed (new IND filed 2001)	1985
	Acambis: (cell-culture vaccinia)	ACAM2000: Licensed 2007	Current
	Bavarian Nordic: (Modified vaccinia Ankara)	Phase II	
Tularemia	Live attenuated vaccine	IND (filed 1965)	1985
Viral Hemorrhagic Fevers Include:	Ebola (NIAID)	IND (Phase I and II versions)	
	Marburg (NIAID)	Phase I	
	Argentine hemorrhagic fever	IND (filed 1985): Phase II	
	Bolivian hemorrhagic fever	IND	
	Eastern equine encephalitis (USAMRMC)	IND (filed 1967): Phase I	1992

Viral Hemorrhagic Fevers Include:	Venezuelan equine encephalitis,		
	Inactivated	IND (filed 1975)	1981
	Live attenuated (USAMRMC)	IND (filed 1965)	1972
	Western equine encephalitis,		
	Inactivated	IND (filed 1984): Phase I	1972
	Sabia-associated hemorrhagic fever		
	Guanarito		
	Lassa fever		
	Lymphocytic choriomeningitis		
	Crimean-Congo hemorrhagic fever		
	Hantavirus		
	Rift Valley fever		
	Inactivated (USAMRMC)	IND (filed 1969): Phase I	
	Live attenuated	IND (filed 1991): Phase I	

Note: The botulism toxoid, tularemia, and viral hemorrhagic vaccines were all developed in military labs and scaled up for pilot production at the Government Services Division of the Salk Institute in Swiftwater, PA.

Source: Institute of Medicine, *Protecting Our Forces: Improving Vaccine Acquisition and Availability in the U.S. Military* (Washington, DC: National Academies Press, 2002).

Key:

Preclinical	Animal model safety and efficacy testing
IND	Investigational New Drug: FDA grants permission to begin Phase I trials in humans after extensive review of preclinical animal data
Phase I	Test vaccine in 20–80 volunteers for safety and immunogenicity
Phase II	Test vaccine in hundreds to several thousand volunteers for immunogenicity, dose ranging, and safety
Phase III	Multicenter trials with several thousand subjects for efficacy, immunogenicity, and long-term safety

licensed just one new vaccine against the Category A biological threats identified by the CDC (an updated version of the smallpox vaccine using cell culture techniques).[52]

Soon after the anthrax attacks, the FDA relicensed a 1970s version of the anthrax vaccine to BioPort (renamed Emergent BioSolutions in 2004). This license does not represent innovative activity, however, because it relies on the same formula developed forty years ago. The failure to license a next-generation anthrax vaccine is particularly disturbing, given that it has been a priority for the Department of Defense since the first Gulf War in 1991.

Several factors drive demand for a new anthrax vaccine. Initially, the Anthrax Vaccine Adsorbed (AVA or BioThrax) was suspected of contributing to a constellation of symptoms associated with the Gulf War Syndrome after it was administered to some troops prior to the 1991 war in Iraq. While multiple studies have since ruled this out, AVA is not suited to emergency use because it required six (now five) shots over an 18-month period, and it was difficult to manufacture.[53] Most importantly, questions remain about AVA's efficacy against weaponized anthrax, although it is not clear that a newer version—the recombinant protection antigen (rPA) vaccine—would be any more effective than the current anthrax vaccine. As with most biodefense vaccines, human efficacy trials would require the deliberate exposure of subjects to lethal pathogens. Given such obvious ethical constraints, neither the smallpox nor the anthrax vaccine has been clinically proven to protect humans against disease from weaponized agents or inhalational challenge.[54]

Although the need for new and improved biodefense vaccines is clear, the means for acquiring them is not. A recent study conducted by the Defense Advanced Research Projects Agency (DARPA) and the University of Pittsburgh Medical Center (UPMC) determined that seventeen novel biologics (eleven of which are vaccines) are currently required to meet military and civilian biodefense needs.[55] This requirement is particularly large when one considers that the FDA licenses an average of only 0.4 vaccines annually.[56] Furthermore, this low innovation rate applies to vaccines with market demand. The innovation rate for biodefense vaccines is far lower because the market pull for this class of biologics is poor at best.

Even when biodefense vaccine development is highly subsidized, it remains an unattractive business for most firms. Unlike seasonal flu or pediatric vaccines, biodefense vaccines have episodic demand from a single buyer—the government. Since they are manufactured for a national stockpile, companies must assume that they will have limited production runs and that expensive facilities will lie idle. As the single buyer, the government also has an unfair advantage in negotiating prices.

Other uncertainties further debilitate the biodefense sector. The FDA introduced the Animal Efficacy Rule in 2002, which allowed researchers to submit animal efficacy data in lieu of human efficacy data. This removed a regulatory hurdle for rare diseases, but it also introduced a scientific challenge to find appropriate animal models for human diseases. Manufacturers have also been uncertain of their intellectual property rights and liability in an emergency. While the first concern remains valid (Chapter 6), recent legislation has alleviated many liability concerns. The 2005 Public Readiness and Emergency Preparedness Act and the Patient Safety and Quality Improvement Act limit liability for companies producing pandemic flu and other biodefense medical countermeasures.

One of the greatest flaws with the federal biodefense development strategy, however, has not been its failure to engage industry, but its preoccupation with stockpiling. There are three serious problems with stockpiling strategies. First, the number of pathogenic threats (natural and manmade) far outstrips our development resources. Development is expensive (approximately $800 million per countermeasure), time consuming (up to ten years), and risky (failure rates reach 50 percent in late-stage development). According to one study, costs may be even higher for biodefense products, reaching $14 billion to develop eight medical countermeasures by 2015.[57]

Second, a stockpile is a fixed defense. As such, it will not deter a technologically advanced opponent. World War II programs demonstrated that it requires far less time, money, and skill to develop a biological weapon than it does to defend against it (Chapter 3). The technological and economic barriers to developing an effective biological weapon are low (relative to nuclear weapons) and getting lower. In 2000, the Defense Threat Reduction Agency (DTRA) demonstrated that, in just over a year, a small group of employees with $1.6 million could buy equipment on the open market to develop biological weapon agent simulants (*Bacillus globulii* and *Bacillus thuringiensis*).[58] A 2006 study by the National Academy of Sciences demonstrated that the growth of the global biotechnology industry is driving down the technological, economic, and educational barriers to weapons development still further.[59]

Third, vaccine stockpiles are not merely impractical, they are hubristic. Future biological threats are likely to be unannounced and unfamiliar, much like the outbreak of SARS in 2002, H5N1 in 2003, and H1N1 in 2009. Furthermore, epidemiological forecasts and threat assessments are notoriously unreliable. Large-scale immunization programs against botulinum toxin in World War II, swine flu in 1976, anthrax prior to the first Gulf War, and smallpox prior to the second Gulf War all addressed threats that failed to materialize.[60] Conversely, military and civilian populations were not adequately prepared for anthrax in 2001, SARS in 2002, or H1N1 in 2009.

Given the multitude of pathogenic threats and the limits of prediction, stockpiling is best restricted to a small set of high-risk pathogens. Smallpox and anthrax, for example, by virtue of their communicability, stability, lethality, accessibility, ease of dissemination, and/or ability to instill public panic could generate enough havoc to qualify as a national security threat. Smallpox and anthrax vaccines can also be used in postexposure scenarios. Beyond this small subset of pathogens, however, defensive, diplomatic, and deterrent strategies are better served by building a system that allows quick reactions to any rapidly evolving or unexpected biological threat, be it natural or manmade.

Given the unpredictable and inevitable evolution of biological threats, it makes more sense to shift from a predictive stockpiling model to a flexible response model that could yield a countermeasure within months rather than years of identifying a novel pathogen. The strategic objective of this model is to catch up to novel threats, even if it is not possible to predict them.[61]

Any genuine attempt to pursue a flexible response model will require biodefense programs to shift their focus from specific countermeasures to the development process itself. These programs should emphasize research tools and component technologies that will shorten the development timeline and/or introduce flexibility into the MCM development process. Examples include rapid detection and diagnostics to speed identification of novel pathogens and diseases, better disease models and biomarkers, DNA vaccine scaffolds, rapid expression systems for injectable proteins, disposable bioreactors for agile facilities, adjuvants to induce immunity more quickly and/or to improve the immunogenicity of DNA vaccines, thermostable formulations, and "no-needle" delivery systems to facilitate the distribution and administration of MCMs in an emergency.

While the goal of this research program is to streamline the development process, it will still be necessary to organize research around pathogen-specific countermeasures. The objective is to develop a range of generic development approaches that can be tailored to specific threats as they arise. Examples include platforms that use viral vectors in cell culture, recombinant proteins, DNA vaccines, and monoclonal antibodies.

The Obama administration has recently proposed a new Center for Innovation in Advanced Development and Manufacturing (ADM) to build rapid and flexible development processes for a wider range of pathogenic threats.[62] ADMs are an important first step toward developing a reactive capability. However, an emergency MCM program must encompass more than late stage development and manufacturing. This program must integrate an entire diagnostic, development, testing, and delivery system to maximize every opportunity to accelerate timelines (Chapter 7).

This program will also require a new regulatory and legal framework to support the development and delivery of emergency MCMs. This new framework will consist of novel validation methods, international agreements to obtain pathogen samples more quickly, harmonized data collection standards to pool clinical data, emergency-use protocols, and liability and compensation agreements. The net effect of these research and development programs, regulatory innovations, and international agreements would be to shorten the time it takes to identify, develop, and administer a safe, effective countermeasure.

Improving our reactive capability is critical. When the United States was confronted with a less familiar virus, like SARS in 2002 and 2003, our MCM apparatus never caught up to the virus. A robust reactive capability would limit the number of casualties from unforeseen natural diseases that move more slowly. Even if vaccines took six months to make, as was the case with the 2009 H1N1 pandemic, health-care workers would still have time to prevent second and third waves of infection. A reactive capability could also contain the damage from less accomplished bioterrorists (i.e., those that have a stolen pathogen in limited supply without the knowledge or equipment to produce more, or those that do not know how to tweak their pathogen to evade countermeasures). Most importantly, a shift from fixed to flexible defenses would terminate a losing battle with an expensive, lengthy stockpiling strategy.

A reactive system will not stop a sophisticated bioterrorist or a fast-moving pandemic, and it may not serve as a powerful deterrent, since attacks would still cause havoc as the United States scrambled to respond. Furthermore, there will always be unforeseen stumbling blocks in the research and development process, and it is unlikely that this program will ever produce "real-time" reaction capabilities. Even under the best of circumstances, vaccine development will be months behind an outbreak, as was the case in 2009.

In many respects, the 2009 H1N1 outbreak represented a best-case scenario for a pandemic: (1) a mild outbreak in the spring of 2009 provided the United States with an early warning about vaccine production needs in the fall; (2) the U.S. government had recently invested in private-sector surge-production capacity following the H5N1 outbreak; (3) influenza virus tracking and vaccine production processes are a relatively well-honed, well-compensated routine; and (4) the government had an opportunity to issue vaccine development contracts at least six months before the fall flu season, giving industry a guaranteed market and clear production benchmarks.

Even so, the first batches of vaccine were not available until October 2009, when H1N1 was already spreading through the population. While several high-risk populations had access to the vaccine at this time, the vaccine did not become widely available until January 2010, after the pandemic had already peaked in the northern hemisphere.

The most startling aspect of the 2009 pandemic vaccine development campaign is that it looked a lot like every other influenza vaccine development campaign since the 1950s, with a six- to ten-month production run determined by the need to cultivate influenza viruses in chicken eggs. Given the remarkable advances in cell culture techniques, basic immunology, and recombinant DNA technology, it is worth asking why twenty-first-century vaccine development campaigns continue to strain under the weight of midcentury methods. The following chapters take a deeper look at the factors that inhibited progress in these critical areas toward the end of the twentieth century.

2
Historical Patterns of Vaccine Innovation

The midtwentieth century is studded with glittering achievements in vaccine innovation, ranging from the development of the first influenza and pneumococcal vaccines in the 1940s to the first polio, measles, mumps, and rubella vaccines in the 1950s and 1960s. By the late 1970s, however, innovation began to taper off and vaccine manufacturers began to exit the industry in droves.

In many respects, vaccines are a public good. Each individual vaccinated for a communicable disease confers a benefit to society by building a firewall of immunity that can interrupt disease transmission to the unvaccinated. Aside from clean water, vaccines are also one of the most cost effective ways to prevent disease. With few exceptions, however, vaccines are manufactured in the United States by the private sector, which is not obligated to provide vaccines when they are needed.

Alarmed by industry contraction in the 1970s, government agencies and think tanks sponsored studies to examine U.S. systems for vaccine innovation and supply. The Office of Technology Assessment (OTA) conducted one of the earliest studies, noting that the number of manufacturers dropped from thirty-seven in 1967 to eighteen in 1979.[1] Of these eighteen remaining companies, only eight were actively manufacturing vaccines for the U.S. market. The total supply of vaccines dropped dramatically during this period, with the number of licensed biological products falling from 385 on the market in 1968 to 150 in 1979.[2] "The apparently diminishing commitment—and possibly capacity—of the American pharmaceutical industry to research, develop, and produce vaccines," they concluded, "may be reaching levels of real concern."[3]

The Institute of Medicine (IOM) followed with another study in 1985, warning that the demand for vaccines was insufficient to support socially optimal levels of innovation. The IOM cautioned, "Our reliance on market incentives to ensure vaccine availability may lead to a failure to meet public health needs. Also, these incentives may not result in optimal levels of vaccine innovation."[4]

The OTA and IOM attributed industry decline in the late 1970s and early 1980s to poor market incentives, citing product liability concerns, rising regulatory costs, and an unfavorable market structure. These factors, they argued, undercut profits from vaccine sales and discouraged industry investment in vaccine research and development. Data provided to the IOM committee from one pharmaceutical company indicated that their vaccine operations, which contributed less than 5 to 15 percent of the company's overall sales, were responsible for 40 percent of all liability claims.[5] Such data left little doubt in the minds of IOM committee members that pharmaceutical companies would favor pharmaceutical over vaccine investments going forward, thus jeopardizing future rates of vaccine innovation.[6]

The IOM asserted that stricter regulations had also raised the cost of vaccine development, which further discouraged industry investment. The 1962 Kefauver Amendments to the Food, Drug, and Cosmetic Act introduced new efficacy and safety requirements for Investigational New Drug (IND) applications. Initially, the OTA noted that these amendments had no visible effect on the number of licensed products and establishments: "During the next five years, the number of licensed products dropped very little, from 396 to 385, and the cumulative number of licensed establishments dropped by two."[7] The new regulations were stipulated in a fragmentary fashion under portions of the Public Health Service Act and the Food, Drug, and Cosmetic Act, and the FDA did not enforce these new standards on vaccines already on the market. In 1972, however, the FDA combined preexisting Public Health Service (PHS) regulations with the 1962 Kefauver Amendments to create a more stringent, uniform set of standards for vaccine safety and efficacy and in 1973, the FDA began to exercise its authority to remove noncompliant vaccines from the market.

Pharmaceutical companies argued that when these regulations were more strictly enforced, costs rose and U.S. companies became less competitive against foreign firms. While it was difficult for the IOM to test the validity of industry's claims without more precise information about their cost structures, the license record suggests that stricter regulations may have had a short-term negative effect on vaccine supply. Between 1973 and 1980, fifty-three vaccine licenses were revoked, a rate significantly higher than the twenty-two vaccine licenses revoked in the prior eight years (1965–1972).[8] However, it is not clear that these declines can be attributed directly to the FDA. Eli Lilly, Dow Chemical, and other large license holders exited the vaccine business during this time period, citing high liability costs, not stricter regulation, as the primary reason.

Rather than inhibiting innovation, some have argued that higher efficacy standards have inspired innovation by allowing the most effective drugs, rather than the most aggressively marketed drugs, to penetrate the market.[9] One study demonstrated that new drug approvals dropped after the Kefauver Amendments came into effect, but that a higher percentage of the total drugs approved had received priority review. The FDA only grants priority to drugs that provide a substantial advantage over existing products. So, even though fewer drugs were approved, one could argue that those that were approved were more innovative.[10]

It is harder to gauge the effect of these regulations on vaccines, partly because vaccines had already been subject to stricter standards. Vaccines, unlike pharmaceuticals, had dual-licensing requirements until 1997.[11] The process of scaling up from pilot lots to large-scale production is seldom linear for biologicals so the FDA required manufacturers to obtain a separate license for their plants to demonstrate that they could produce a safe, effective vaccine on a commercial scale before the vaccine entered large-scale clinical trials. Vaccine developers, therefore, had to make heavy up-front capital investments in manufacturing facilities to obtain an establishment license before they knew if they had a viable product.[12] Because developers had to invest in manufacturing facilities before a product license was assured, these investments generated a lower internal rate of return than pharmaceutical plant investments, which could be made after FDA approval.[13] Maintenance costs are also higher for vaccine plants because biological outcomes are more sensitive and temperamental than chemical manufacturing processes.

While dual-license requirements were not new, other factors drove up the overall cost of vaccine clinical trials during this period. Prior to the 1970s, it was not uncommon to test vaccines in institutionalized residential settings, such as military barracks, prisons, and homes for mentally disabled children and adults. Close living quarters bred high disease rates in a controlled setting. These conditions allowed researchers to demonstrate vaccine efficacy quickly, efficiently, and inexpensively. Abuses of the system, however, led to large-scale reforms in clinical trial protocols. In one well-known case in the 1960s, Saul Krugman deliberately infected children at the Willowbrook School with hepatitis to learn how the virus was transmitted. As result of this and other historical abuses (i.e., the Tuskegee syphilis study), in 1974 Congress passed the National Research Act, which mandated human subject research regulations. These reforms dramatically increased the paperwork that companies needed to file with investigational review boards and ethical committees for human-subject testing. Over time, vaccine safety requirements became more stringent as well and clinical trials grew larger, longer, and more expensive. In 1980, the FDA updated the standards for Good Manufacturing Practices (GMP), raising the cost of vaccine manufacturing still further—by $200 million according to some estimates.[14] Chronic regulatory updates to the production process have continued to raise costs, extend production times, and limit the advantages of manufacturing learning curves.

Regulatory disincentives aside, economists have established that commercial markets often fail to produce socially optimal levels of vaccine innovation and consumption.[15] This effect is most easily observed in markets for global health vaccines and biodefense vaccines. Demand for these vaccines falls well short of desirable levels of innovation and supply. Economic disincentives exist even for vaccines with viable markets. One such disincentive is inherent to the nature of vaccines themselves; products that offer long-term immunity after a minimal dose confound most profit-making strategies, since pharmaceutical companies prefer high-volume repeat sales. Some vaccines, used effectively, can eliminate their own market by eradicating the disease they prevent, as was the case with smallpox.

Consumer behavior further weakens vaccine markets. People are more willing to pay to treat acute and existing conditions than they are to prevent an uncertain condition that might arise in the future. This is true even if the former condition is temporary and the latter condition is life threatening. As one industrial research scientist explained: "Curative measures are more profitable in general because demand is always highest in the face of a salient illness. The need for curative measures is more chronic than outright prevention, which kills its own market after the first few immunizations. The vaccine industry puts itself right out of business."[16] Most consumers also do not account for the societal benefits of a vaccine when they consider what price they would be willing to pay for it.

The structure of the market for vaccines also weakens incentives for private-sector investment. Governments are often the largest buyers of vaccines. The Federal Centers for Disease Control and Prevention (CDC), for instance, accounts for over 50 percent of all pediatric vaccine sales in the United States. As the largest single buyer, the government exercises monopsony power to negotiate lower prices. This ability to drive down prices was apparent in the American Enterprise Institute (AEI) report, which observed that the ratio of public to private prices for major childhood vaccines, such as the combined diphtheria, tetanus, and pertussis (DTP) vaccine and the combined the measles, mumps, and rubella (MMR) vaccine, ranged from 25 percent to 54 percent in 1996.[17] Even though the cost of individual vaccines has gone up over the years, public purchase prices are still significantly lower than the prices charged in private markets. On average, private prices are 42 percent higher for pediatric vaccines and 58 percent higher for adult vaccines.[18]

Industry consolidation may have stifled innovation as well. Whereas the vaccine market had once been populated with dozens of small, dedicated vaccine manufacturers, most vaccines today come from five pharmaceutical conglomerates: Merck, GlaxoSmith-Kline, Sanofi-Pasteur, Pfizer, and Novartis. As stand-alone vaccine producers, executives chose among vaccine projects; as conglomerates, executives often considered the opportunity costs associated with vaccine investments against the blockbuster potential of pharmaceuticals.[19] The annual sales of a single cardiovascular drug like Lipitor, for example, ($11.4 billion in 2009) amounted to half of the annual returns for the entire global vaccine industry ($22.1 billion in 2009). Consolidation also reduced competition among firms, further limiting incentives to invest in vaccine innovation.[20]

Since the vaccine industry consolidated in the 1980s, supply shortages have been a perennial problem.[21] Consolidation improved profit margins by creating product-line monopolies, but it left the vaccine supply more vulnerable to production problems within individual companies. Whenever a plant experiences problems with a production run, the developer often must restart the process. If they are the sole producer, supplies can be disrupted for up to a year before the company is able to grow and purify enough antigenic material.

In February 2002, after several companies experienced production problems, the CDC announced that there were temporary shortages of eight out of eleven of the most commonly administered childhood vaccines.[22] The director of the CDC's immunization program stated that such wide-scale shortages were "unprecedented—posing a threat to the public health that's far more concrete than the fears that led so many to panic about anthrax and smallpox."[23]

The OTA and IOM wanted to demonstrate not only that supply was becoming more fragile but that innovation was declining as well. The OTA asserted that vaccine innovation rates had suffered, but it could not demonstrate that the number of new vaccine product licenses had declined over time.[24] Their data indicated that innovation rates were holding steady. The OTA report argued instead that the dramatic reduction in the number of vaccine manufacturers would eventually cause innovation to suffer. The IOM was also unable to demonstrate a decline. On the contrary, using the same type of data as the OTA study (i.e., official lists of currently licensed vaccines), the IOM demonstrated rising rates of innovation.

Why were these studies unable to find evidence to support the widespread perception that innovation rates were declining through the 1970s and 1980s? A rigorous review shows that the data used in both studies were deeply, if inadvertently, flawed.

The name and/or location of the institution responsible for regulating biologics has changed seven times since the first licenses were issued in 1903 (Table 2.1). Each regulatory body maintained an accurate record of current vaccines, but did not maintain records for companies that left the industry or for licenses that had been revoked. Each time regulatory responsibility changed hands, original approval dates were either lost or reentered with a more recent date, creating the false impression that there was a spate of innovation with each transition. Compounding this error, in the 1970s the FDA began to issue licenses for vaccines that were new in name only or for license transfers between companies. These superficial name

Table 2.1 Institutions Responsible for Vaccine Regulation

Date	Institution	Regulatory Body
1903–1930	Public Health Service Hygienic Laboratory	PHS
1930–1937	Hygienic Laboratory	NIH
1937–1948	Laboratory of Biologics Control	NIH
1948–1955	National Microbiological Institute	NIH
1955–1972	Division of Biologic Standards	NIH
1972–1982	Bureau of Biologics	FDA
1982–1987	Center for Drugs and Biologics	FDA
1987–Present	Center for Biologics Evaluation and Research	FDA

changes and license transfers do not represent innovative activity, yet they account for an overwhelming 64 percent of the new licenses issued since 1970 (Figure 2.2).

This simple misunderstanding of vaccine license data biased innovation studies. The OTA based its analysis on the BOB's list of currently approved vaccines in 1979, and the IOM relied on the FDA's list in 1983. Neither list accurately recorded original approval dates and by 1983, the FDA list was missing more data than the BOB list had been. The OTA listed forty-nine vaccine licenses issued since 1903 and found no significant changes in historical rates of innovation.[25] The IOM listed only thirty-four significant vaccine introductions since 1903. Even though the IOM recorded fewer introductions, their study suggested that innovation rates had increased considerably since the 1940s, simply because they underreported historical rates of innovation (Table 2.2).

Table 2.2 Historical Patterns of New Vaccine Product Introductions

	OTA (1979)[a] BOB, "Currently Licensed Vaccines in 1979"	IOM (1985)[b] FDA, "Currently Licensed Vaccines in 1983"	Historical Record of Vaccine License Data[c]
Before 1940	10	Not recorded	11
1940–1949	9	4	11
1950–1959	10	2	9
1960–1969	10	10	10
1970–1979	10	14	8
1980–1989		4 (1980–1983)	2
1990–1999			5

a. Office of Technology Assessment, *A Review of Selected Federal Vaccine and Immunization Policies: Based on Case Studies of Pneumococcal Vaccine* (Washington, DC: U.S. Government Printing Office, 1979), 31. See Table 5.

b. Institute of Medicine, *Vaccine Supply and Innovation* (Washington, DC: National Academies Press, 1985), 50. See Table 4.3.

c. Appendix 1.

When one applies OTA/IOM methodologies for selecting new vaccine licenses to a more accurate dataset (i.e., one that includes original approvals and excludes licenses for noninnovative activity), clear differences in innovation patterns emerge. Table 2.2 demonstrates that innovation rates have not held steady (OTA study) or risen (IOM study), but have, in fact, fallen.

Biases arising from inaccurate historical data are also evident in more recent studies. The AEI asserted that the vaccine industry experienced a "renaissance" in the 1980s and 1990s.[26] Without clear data to support their claims, the authors used individual anecdotes to extrapolate industrywide trends. They cited individual cases of new vaccine development since the late 1980s: *Haemophilus influenza* type B, hepatitis A and B, varicella, and acellular pertussis.[27] The authors also cited rising research and development investments as evidence.[28] They indicated that the portion of company research and development budgets devoted to biologicals fell from 5 percent in the early 1970s to less than 2 percent in the early 1980s but that investments rebounded after 1986 from 4 percent to 5 percent in the early 1990s.[29] These increases, they concluded, reflected new technological opportunities in genetic engineering and peptide chemistry and improved economic incentives following the passage of the 1986 National Childhood Vaccine Injury Act, which created a no-fault system for vaccine-related injuries.

While larger investments may have reflected optimism, they did not translate into a higher number of new vaccines relative to previous decades. In reality, innovation rates continued to fall through the 1980s and 1990s (Figure 2.1).

TO OBTAIN AN ACCURATE picture of innovation patterns, I constructed a new database that contains the original approval dates for new vaccines. Because original materials were often discarded with each transfer in regulatory authority, a reliable central repository for historical vaccine license data does not exist.[30] Through the Freedom of Information Act (FOIA), I obtained a partial list of all vaccines licensed since 1903 from the Center for Biologics Evaluation and Research (CBER). Unlike the FDA's list of currently licensed vaccines, this list includes some vaccines that are no longer licensed or marketed in the United States. CBER warned, however, that they could not guarantee the accuracy of original vaccine approval dates before 1987, when they began to maintain their own database.[31]

I drew from a wide range of primary and secondary resources to restore obsolete licenses and to obtain accurate dates for existing licenses. These included historical PHS publications, company product records, correspondence in industry and federal archives, and historical anthologies of vaccine development.[32] I checked the accuracy of these historical records and CBER's data against less detailed information released in publications issued from each agency responsible for the regulation of biologics over the past century. Through these methods, I was able to correct several dozen entries on CBER's official dataset and to restore 99 license entries that had been lost over the years in agency shuffles. CBER's list contained 239 vaccine licenses for 1903–1999, whereas my research yielded a final list of 338 licenses for this period (Appendix 1).[33]

Figure 2.1 uses this new dataset to illustrate the best existing approximation of vaccine innovation patterns in the twentieth century.[34] This figure includes licenses for new or improved vaccines and excludes licenses issued for noninnovative activity, such as superficial name changes, license transfers associated with industry consolidation, and minor changes to the indication, dosing, or formula of a vaccine.[35] It is important to note that while this figure reflects technological innovation, it is not a direct measure of scientific discovery. In many cases, vaccine licensures lagged decades behind the scientific breakthroughs that enabled their development. In this respect therefore, Figure 2.1 reflects industry's ability to translate discoveries into licensed vaccines.

Figure 2.2 illustrates the discrepancy between innovative and noninnovative activity in the latter half of the twentieth century. Licenses representing noninnovative activity form a significant portion of the licenses issued since the 1970s for several reasons. Due to heavy industry consolidation during this period, many licenses for

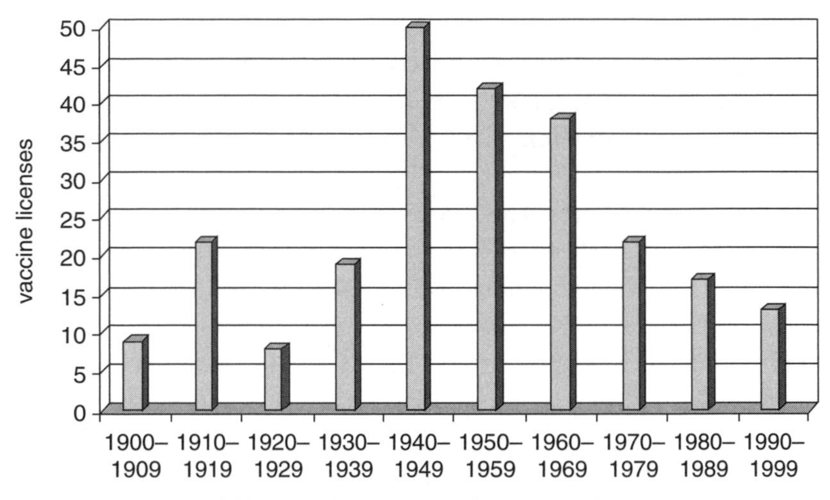

Figure 2.1. Licensed Vaccines Representing Innovative Activity

previously developed products were transferred to new or merged companies, with each transfer resulting in a "new" license for the new owner. Further, when the FDA assumed responsibility for the regulation of biologics from the National Institutes of Health (NIH) in 1972, they lost many original approval dates and entered newer approval dates for older vaccines. Finally, in the 1980s and 1990s, the FDA began to issue product license supplements for minor changes to the formula, dosing, or indication for a vaccine. Together, these recording practices account for the apparent burst of innovative activity in the latter half of the twentieth century.

This new data set yields a very different picture of twentieth-century innovation. Contrary to earlier studies, Figure 2.1 demonstrates that innovation peaked in the 1940s. While innovation remained relatively robust in the 1950s and 1960s, it has steadily declined, with the largest drop occurring in the 1970s. These findings support the original concerns that motivated the OTA and IOM reports and counter claims that the vaccine industry experienced a rebirth of innovative activity in the 1980s and 1990s. More importantly, these results do not support traditional market-based explanations of vaccine innovation.

Previous investigations of vaccine innovation and supply have fo-

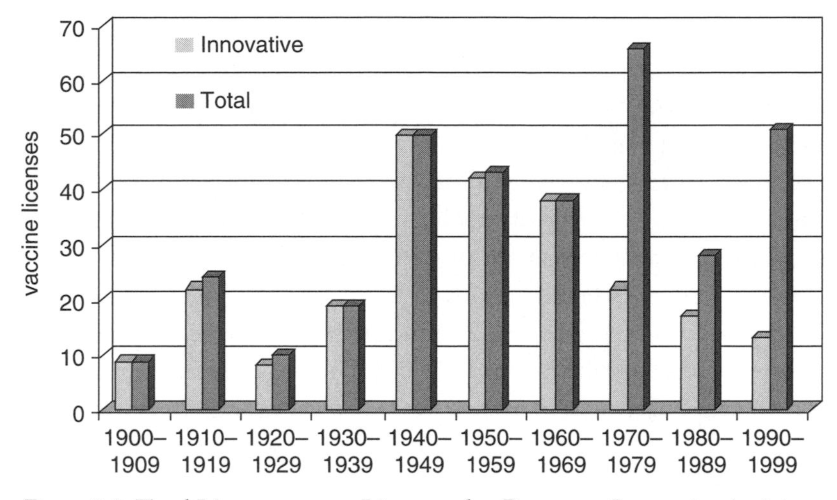

Figure 2.2. Total Licenses versus Licenses that Represent Innovative Activity

cused on the role of market factors.[36] Historians argue that technological innovation is a function of economic incentives, individual firm capabilities, and the available stock of "scientific knowledge" or "technological opportunities."[37] Recent investigations of vaccine development also interpret innovation patterns with reference to the profit-maximizing behavior of commercial vaccine developers.[38] The vaccine innovation patterns presented in Figure 2.1, however, do not submit to prevailing theories on the sources of innovation in industrial settings. Three key innovation factors—incentives, capabilities, and knowledge/opportunities—were weak for the vaccine industry in the 1940s and 1950s, when innovation rates were high, and stronger in the 1980s and 1990s, when innovation rates were low.

The market rationale for vaccine development was at least as unattractive in the 1940s as in the 1980s. The long-term outlook for the vaccine market looked particularly grim after the successful introduction of sulfa drugs in the 1930s to treat bacterial infections. Developers feared that new antiviral medications would not be far behind and that vaccines would go the way of the buggy whip.

Monopsony power dynamics existed during this period as well

because government purchases for military vaccines dominated the market.[39] According to wartime records from the National Drug Company, nearly 100 percent of its output supplied the armed forces during World War II.[40] After a brief slump in government vaccine purchases immediately after World War II, the government became the largest single buyer of vaccines again during the Korean War, accounting for 50 percent of all sales at National Drug. Large government contracts brought sales to an all-time high of $2 million in 1951, yet the low prices negotiated for these contracts diminished margins considerably. By 1953, A. B. Collins, president of National Drug, acknowledged that net earnings were falling despite record sales.

It appears that many companies maintained vaccine divisions during this period, not entirely for economic reasons, but partly out of a sense of public and patriotic duty. One industrial research scientist observed that "all pharmaceutical companies had a vaccine business and looked at it as a public service by the company, not as huge revenue generators, which they weren't."[41]

In many respects, the long-term economic outlook for vaccines improved dramatically toward the end of the twentieth century. By the early 1980s, profits began to improve. Don Metzgar, a vaccine research scientist at Aventis Pasteur (formerly Connaught) during this period, observed, "Ironically, Connaught began to make more money on vaccines when litigation became so prevalent. We couldn't get insurance because the price had gotten so high . . . Connaught started to charge more money to cover self-insurance needs. There was a period of time when we didn't sell anything—almost drove ourselves out of the market. Eventually, other companies began to look at what we were doing and followed suit."[42]

Pricing data from that time period supports Metzgar's observation. Since 1985, vaccine prices have been rising in real terms at a faster rate than consumer and producer price indexes.[43] Data provided by the CDC demonstrate that vaccine prices, relatively low and stable through the 1970s, jumped considerably in the late 1970s/early 1980s and continued to rise through the 1990s.[44] The price of a common pediatric vaccine, DTP, for example, jumped 2,847 percent between 1977 and 1991.[45]

Vaccine prices continued to grow through the 1980s and 1990s as well. In 1986, the government placed a surcharge on the sale of all vaccines to create a fund to manage liability claims. This program removed the need for industry to self-insure through price hikes, but companies continued to raise their prices.[46] The cost of immunizing a two-year-old doubled from 1999 to 2002, as higher-priced new childhood vaccines—such as the varicella and conjugate pneumococcal vaccines—were added to the recommended immunization schedule.[47]

The competitive landscape for vaccine production changed as well. Consolidation gave surviving firms a greater controlling share of the market. The four largest remaining firms (at the time)—Merck, Wyeth, Aventis Pasteur, and GlaxoSmithKline—went from controlling 50 percent of the U.S. market in 1988 to controlling 75 percent to 80 percent in 2001.[48] Several of these firms developed monopolies in noncompeting product lines. Merck, for example, became the sole supplier of MMR and varicella vaccines. Wyeth supplied the DTP and conjugate pneumococcal vaccines, while Aventis Pasteur supplied the meningitis, polio, and yellow fever vaccines.

Consolidation provided a countervailing force to government monopsony power in price negotiations. In 1987, federal contract prices were 75 percent less than catalogue prices; by 1997, they were only 50 percent less.[49] The overall size of the vaccine market was growing as well. According to one study, the global vaccine market increased 10 percent a year during the 1990s.[50]

The long-term need for vaccines was becoming increasingly apparent as well: nearly eighty new infectious diseases had emerged (and reemerged) since the 1970s, antibacterial resistance had grown, few effective antivirals had been developed, and biosecurity threats had become more salient (Chapter 1). As generic competition grew and research and development productivity fell, manufacturers have begun to regard growth opportunities in the vaccine sector with greater interest. New funding sources (through the Gates Foundation) and novel financing mechanisms (such as advance market commitments) have also begun to fill development pipelines and expand markets.

The vaccine industry has enjoyed exceptional technological opportunities in the late twentieth century as well. In the 1940s, research scientists relied on imprecise and inefficient methods to grow viral cultures for vaccine production. Viral cultures for the smallpox vaccine were grown and scraped off cow bellies, for example, whereas now researchers can more safely and reliably grow viruses in cell cultures. Midcentury bacterial fermentation techniques were also temperamental and unsophisticated by today's standards. By the 1980s and 1990s, advances in molecular biology and genetic engineering created new opportunities for precision and purity, permitting scientists to engineer immunogens apart from the whole pathogen and thereby develop safer vaccines.

To capitalize on these opportunities, the pharmaceutical industry was investing a greater portion of its research and development budget in biologicals. Before World War II, few vaccine developers set aside significant earnings for in-house research, and some commercial firms maintained no more than five scientists on staff. Since the 1940s, however, in-house research and development capabilities have grown dramatically and vaccine developers now employ hundreds of research scientists.

As the biotechnology revolution gained momentum in the 1980s, the number of small, dedicated biotechnology firms grew from a handful in the 1970s to 1,457 in 2001.[51] Larger established firms also began to form partnerships with academia and to build their own capabilities in molecular biology and genetic engineering. The pharmaceutical industry's research and development budget for biologicals rose from 2 percent to 4 percent of total spending between 1985 and 1990.[52]

Even so, greater economic incentives, technological opportunities, and firm capabilities did not correlate with higher rates of vaccine innovation during the latter half of the twentieth century.[53] One could argue that innovation rates had fallen because midcentury developers had already picked the low-hanging fruit, leaving only those vaccine candidates that posed larger scientific and technological challenges to be developed. The protracted struggle to develop an

HIV vaccine offers strong evidence for this argument. While HIV vaccines pose exceptional scientific and technological challenges today, it is worth recalling that many midcentury vaccines were not considered easy targets at the time either. The first pneumococcal vaccine (licensed in 1948) was a radical innovation by any standard since it was the first successful use of capsular polysaccharides to generate an immune response. Other momentous achievements included the first use of eggs to cultivate viruses for yellow fever and influenza vaccines (the 1930s and 1940s), and the use of kidney tissue cultures to cultivate polioviruses (the 1950s).

The low-hanging fruit argument also fails to explain why vaccine developers have failed to "pluck" many of the technologically feasible vaccines languishing in the pipeline today, such as the recombinant protective antigen anthrax vaccine. Most of these vaccines stall in late-stage development, not for scientific or technical reasons, but for a collection of organizational and financial reasons discussed in chapters 5 and 6.

In sum, traditional market-based formulas for innovation in industrial settings predict that greater economic incentives, technological opportunities, and firm capabilities in the latter half of the century should have led to a corresponding rise in innovation rates. Yet the data demonstrate a decline. These findings suggest that a wider range of factors influenced innovation patterns in earlier decades. Which factors have not been considered, and when did these factors trump traditional market factors?

To answer these questions, I surveyed the developmental history of vaccines licensed in the United States during the twentieth century. In each case, I identified the original license for a new vaccine, the developer, and the type of innovation each new license represented (Appendix 2).[54] Scientists developed vaccines for twenty-eight diseases in total between 1903 and 1999. These new vaccines hailed from a wide range of institutional settings (public, private, and hybrid arrangements), but a few stand out for the number and quality of their contributions. Military research institutions made significant contributions to eighteen of the twenty-eight vaccine-preventable diseases (Appendix 3), and Merck & Company made extensive contributions to ten of these. While the talents of specific scientists were critical, the presence of innovation clusters suggests that institution-based research practices and collaborative networks played an important role as well.[55]

By prying into the developmental history of these vaccines, I was able to move beyond strict input (research and development budget, personnel, lab space) and output analyses (number of new products) to examine what happens inside the development process itself. This historical approach illuminates our understanding of vaccines for three main reasons. First, decisions about whether to develop vaccines, and which vaccines should be developed, are not wholly determined by the market. Historically, other institutions (such as the military) have prevailed on firms to develop and license vaccines, even when market conditions were poor.[56] Second, vaccine development is a highly interdisciplinary endeavor that requires contributions from a wide variety of experts and institutions. The firm obtaining the license, therefore, is never the sole source of innovation; the incentives and contributions of collaborative partners must be taken into account as well. Third, vaccine development requires the manipulation and growth of microbial and/or mammalian cells, activities that do not always yield predictable results. The development process is laden with tacit knowledge that can be difficult to transfer from lab to lab or person to person. To understand why certain projects succeed or fail, therefore, it is useful to examine working relationships among collaborators.

This historical approach sheds new light on the research practices, institutional networks, and working relationships that drove innovation. Viewed through this lens, it appears that a collection of nonmarket factors had a strong influence on twentieth-century innovation patterns. While these practices were often employed by market-driven firms, they were not always in themselves a response to market pressures, but a reflection of the ideologies, habits, politics, and values of a particular community of vaccine developers.

3
Vaccine Development during World War II

War and disease have gone hand in hand for centuries. As one historian has observed: "More than one great war has been won or lost not by military genius or ineptitude, but simply because the pestilence of war—from smallpox and typhoid to cholera, syphilis, diphtheria, and other scourges—reached the losers before they infected the winners."[1]

Training camps and battlegrounds magnify the spread and severity of disease.[2] They bring men from different geographical regions into close contact with one another. These men are often physically stressed, or wounded, and disease spreads easily. Prior to World War II, soldiers died more often of disease than battle injuries.[3] The ratio of disease to battle casualties was two to one in the Civil War and approximately five to one during the Spanish-American War.[4] Severe losses, from typhoid fever in particular, inspired the U.S. Army to sponsor the research of Major Fredrick Russell, who succeeded in developing an effective typhoid fever vaccine for the military in 1909.

Improved sanitation measures lessened disease casualties in World War I, but failed to protect troops from the 1918 influenza pandemic. Military populations were particularly hard hit. According to some estimates, influenza accounted for nearly 80 percent of the war casualties suffered by the U.S. Army during World War I.[5]

The 1918 influenza pandemic first appeared in the United States at an army training camp at Fort Riley, Kansas, in March, sickening hundreds. By September, the flu spread to Camp Devens, Massachusetts. By October, it crossed the country, infecting recruits at the University of Washington Naval Training Station. Troop movements facilitated global transmission, contributing to three near simultaneous outbreaks in the port cities of Boston; Brest, France; and Freetown, Sierra Leone, in that same year. The flu spread rapidly from these port cities, claiming approximately fifty million lives worldwide.[6]

Thomas Francis, Jr., professor at the New York University College of Medicine and chairman of a commission that coordinated research on the influenza vaccine during World War II, feared that the war would generate the epidemiologic conditions for another pandemic. He remarked: "The appalling pandemic of 1918 in the last months of the exhausting conflict of World War I, with massive mobilization of armies and upheaval of civilian populations, has irrevocably linked those two catastrophes. It demonstrated that virulent influenza may be more devastating to human life than war itself."[7]

Fearing another pandemic, war planners began to make unprecedented investments in infectious disease research and vaccine development through intramural and extramural projects conducted through the Office of Scientific Research and Development (OSRD) and the Army Surgeon General's Office (SGO).

Military planners were also concerned about biological warfare. They reasoned that "the devastation wrought by the natural partnership of war and pestilence has scarred the face of history so deeply that it is only logical that military men in search of offensive weapons should consider the intentional use of disease-producing agents."[8]

Often war planners did not distinguish between the threat of natural and intentional forms of disease, in part because the distinction was difficult to make. The influenza pandemic of 1918, for example, had been so severe, so unprecedented, and so devastating to U.S. troops in particular, that Office of Strategic Services (OSS) intelligence reports suggested that the Germans had deliberately unleashed the disease.[9]

James Simmons, chief of the Preventive Medicine Division in the SGO, was an early proponent of programs to defend against biological weapons. While stationed in Panama in 1934, he became so "impressed with the hazard of yellow fever and its possible intentional introduction that he prepared an informal plan to counteract such a move in the event of war."[10] In January 1941, Simmons recommended mandatory vaccinations for all servicemen in tropical stations. Biodefense planning gained momentum as a series of intelligence reports fed anxieties that Axis nations were investing in biological warfare capabilities.[11] Although intelligence officers followed German activities intently, initial reports indicated that the Japanese had a far greater interest in biological weapons. According to one account, a Japanese doctor had attempted to acquire yellow fever virus from the Rockefeller Institute in New York in 1939. When he failed to obtain it by request, he purportedly attempted to bribe an employee.[12] There was also evidence that the Japanese had trained over two thousand parachute troops as a "bacteriological warfare battalion."[13]

In the summer of 1941, intelligence sources implicated the Germans in biological warfare activities. According to an OSS report known as the Bern Report, Professor Menk from the School of Tropical Medicine in Hamburg had been trying to weaponize botulinum toxin in a lab outside of Paris. While the Bern Report contained several widely recognized inaccuracies, military planners required little evidence amid the fear and suspicion surrounding German activities prior to the war.[14] An in-house history of the SGO refers to the Bern Report as an "opportune piece of intelligence which fell into hands well prepared to receive it."[15] The history pointedly notes: "Planted on the fertile soil of the interest shown to the subject by the Surgeon General's Office and Chemical Warfare Service, it appears as if the MA Bern Report took root, flourished exceedingly and formed the excuse, as it were, for the events that followed."[16]

Despite mounting evidence of Japanese interest in biological weapons (BW), it was not until the Germans were implicated that the SGO began to clamor for BW research programs. Simmons sent a memo to Harvey Bundy, special assistant to the secretary of war, demanding that "serious consideration should be given to the advisability of developing facilities within the Medical Department for intensive research on methods for preventing diseases of man, lower animals, or plants that might be introduced artificially by military enemies."[17] The U.S. surgeon general, Thomas Parran, also began to alert civilian officials to the threat of biological warfare. Addressing a mayors' conference in 1942, Parran presented the threat of biowarfare in stark terms, asserting, "The enemy has planned and, in my opinion, will use bacteriological warfare whenever possible." Such tactics, he warned, "can be as deadly as mustard gas or explosives." He urged the mayors "to begin at once to take every possible precaution and to get expert advice."[18]

Academic experts were less convinced of the biological warfare threat. In October 1940, Vannevar Bush, chairman of the National Defense Research Committee, and Lewis Weed, chairman of the Division of Medical Sciences at the National Research Council (NRC), asked the Health and Medical Committee of the Council of Medical Defense if the nation was prepared to handle a biological attack.[19] The council, in turn, asked the National Institute of Health (NIH) and the National Academy of Sciences (NAS) to study the issue and evaluate the threat of biological warfare. An NIH committee of special consultants—mainly university scientists—concluded that "biological warfare was not considered practicable or as constituting a menace to the country."[20]

Dissatisfied with the initial NIH expert assessment, the secretary of war, Henry Stimson, asked Weed to appoint another group to assess the threats and opportunities posed by biological warfare. Weed formed a joint committee of members from the NRC and the NAS, known as the War Bureau of Consultants (WBC), comprised of Alfred Newton Richards, chairman of pharmacology at the University of Pennsylvania; Lieutenant Colonel Jacobs, GSC; Colonel M. E. Barker, CWS; Colonel Simmons; R. Harrison, chairman of the NRC; and Weed. By the time this committee convened, however, the question was no longer whether the United States should proceed with a BW program, but rather, who would be responsible for it.

After the bombing of Pearl Harbor in December 1941, military and civilian officials launched into full-scale offensive and defensive preparations for biological warfare. Military planners shared Francis's concern that a new war would unleash another influenza pandemic, along with a host of other diseases, and the government began to invest in large-scale, federally coordinated vaccine development programs.

Intramural and extramural research projects were administered through the U.S. Army's SGO, the Committee on Medical Research (CMR) division of the Office of Scientific Research and Development (OSRD), and, in some cases, the War Research Service (WRS).[21]

The Army's Preventive Medicine Division directed in-house research programs, using a network of international laboratories and the Army Medical Graduate School (AMS) in Washington. Prior to the war, these programs were chiefly concerned with the diagnosis, prevention, and treatment of typhoid dysentery, typhus, and syphilis. In 1941, Stimson created the Board for the Investigation and Control of Influenza and other Epidemic Diseases (BICIED).[22] Administered through the Preventive Medicine Division, this seven-member board directed ten commissions and one hundred civilian scientists.

These commissions enlisted the top infectious disease specialists in the country—from universities, hospitals, public health labs, and private research foundations—to conduct epidemiological surveys and to develop and test preventive measures against diseases of military importance such as influenza, meningitis, encephalitis, acute respiratory diseases, measles, mumps, pneumonia, typhus, and rickettsial diseases.[23] With the exception of the Respiratory Disease Commission Laboratory at Fort Bragg, North Carolina, the War Department contracted these scientists to conduct research at their home institutions on a part-time basis. According to Stanhope Bayne-Jones, deputy chief of the Preventive Medicine Division during the war, BICIED contracts were designed to permit the army to outsource research they were not qualified to perform, while gaining access to "valuable services and facilities in the leading institutions in the country."[24]

The CMR also contracted civilian scientists to conduct vaccine research for the military.[25] The CMR was created to tap expertise within the Division of Medical Sciences (DMS) at the NRC, since the NRC was not a government agency and by law could not obtain congressional funding to administer a large-scale contract research program. The CMR was chaired by Richards and administered by three presidential appointees and one representative each from the offices of the secretary of war, the navy, and the Federal Security Agency. Lewis Weed, chairman of the DMS, was elected vice chairman of the CMR. The directors of individual DMS committees were appointed as consultants to the CMR.

Members of the organizing WBC committee agreed from the start that defensive and offensive biological programs must be kept nominally, if not institutionally, separate. The SGO would administer defensive research programs and the WRS would direct offensive research. The WBC committee suggested that research on biological weapons should be governed by a civilian agency with the harmless title of the War Research Service (WRS). While George Merck, CEO of the pharmaceutical company that bears his name, agreed to direct the WRS, institutional responsibility for the WRS was passed around like a hot potato. Vannevar Bush, director of the newly formed OSRD, recalled many years later that "Mr. Stimson did not want this thing in the War Department, and I did not want it in OSRD. So we inserted it in an agency headed by Paul McNutt. I don't think Mr. McNutt knew he had it."[26] Thus, as Merck recalled, biological warfare research was taken "under the wing and the cloak of the Federal Security Agency."[27]

Offensive BW research was hidden in a civilian research service during the first half of the war "for the purposes of security and for protecting the armed services from public involvement in biological warfare."[28] The newly founded WRS coordinated the activities of the military, Public Health Service (PHS), Department of Agriculture, FBI, and OSS that pertained to biological warfare, while removing responsibility for these activities from those jurisdictions. The service formed two civilian advisory committees to consider the offensive and defensive aspects of biological warfare. According to a WBC committee report, they "would in effect be one, but . . . separate reports would be prepared for transmittal to the OSG and the CWS, respectively."[29]

Maintaining strict separation between offensive and defensive projects proved difficult. A WRS memorandum regarding U.S. biodefense programs acknowledged that "it is impossible, however, to visualize all the possibilities inherent in bacterial warfare if the question is considered purely from the defensive standpoint. The assumption of an offensive viewpoint is absolutely necessary."[30] The memo concluded, "It is believed that such warfare is of sufficient importance to warrant asking the help of the OSRD."[31]

Thus offensive and defensive research objectives were entwined from the start, and biowarfare concerns motivated a number of vaccine development projects, such as the development of yellow fever, plague, botulinum toxoid, and typhoid vaccines.

A sense of urgency was evident outside of the government as well, as federal officials received numerous offers of assistance from industry and academia. Richards, chairman of the CMR, which coordinated many vaccine development projects under OSRD, marveled at the "unselfish zeal, co-operative spirit, and the competence with which our civilian investigators, laying aside more agreeable pursuits, entered into the attack on problems whose solution was vital to our fighting forces . . . Never before, we believe, has there been so great a coordination of medical scientific labor."[32]

A collection of influential scientists and industrialists had already begun to lay the groundwork for more cooperative business-government relations in support of science during the New Deal era. Individually and collectively, men such as Karl Compton (president of MIT), Gerard Swope (president of General Electric), Isaiah Bowman (chairman of the National Research Council), and Frank Jewett (president of the National Academies of Science and vice president of the American Telephone and Telegraph Company) believed that corporations should support national planning goals.[33] These corporate liberals reacted, in part, to the charge that advances in science and technology had hastened the Great Depression by displacing factory workers from jobs. By taking proactive steps to work with govern-

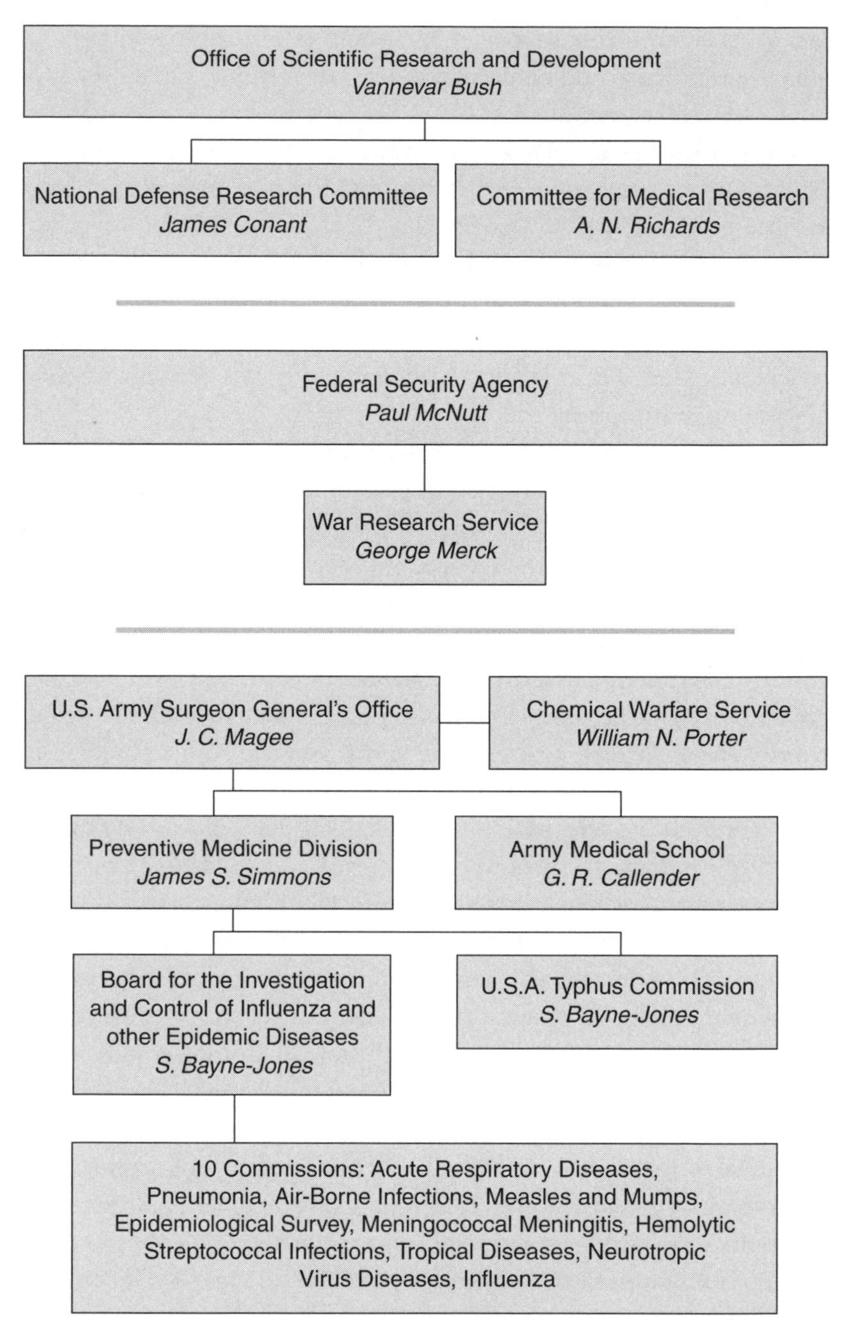

Figure 3.1: Simplified Organization Chart of Military and Civilian Agencies Associated with Vaccine Development, 1941–1942

ment agencies, corporate liberals hoped to both preempt stricter federal control and to demonstrate that science and technology were not the source of—but the solution to—the nation's problems.

Many ideological and practical barriers to closer cooperation eroded in the face of war. The Roosevelt administration made probusiness concessions (such as accelerated amortization measures) to encourage industry to make capital investments in strategically important sectors. Industry also became more willing to convert production lines over to mobilization efforts and to accept low margin contracts as the war intensified. Academic opposition to a technocratic reorganization of the nation's research and development apparatus softened as well and universities redirected the activities of their best researchers to government programs.

Remaining opposition to these cooperative arrangements was tempered by an implicit understanding that all provisions would be temporary. Irvin Stewart, deputy director of the OSRD, believed this short-term mentality accounted for much of the organization's success. He wrote, "The organization was built on a temporary basis, drawing upon the best available men for relatively short periods of time without disturbing their regular academic or industrial connections in most cases. This was possible largely because of the pressure of impending and actual war, which made men available whose service could not have been obtained on any comparable scale in normal times. The leaders of OSRD were always keenly conscious of this fact, which however completely escaped many people on the outside who, seeing the success of OSRD, called for its retention in peacetime. This could never be done."[34]

From the outset, there were signs that intense, coordinated effort could not be sustained. Research and development contracts were issued on a "no loss, no gain" basis. While these contracts covered the cost of research to the performing institution, and occasionally calculated overhead to cover indirect costs of research, they were not designed to be profitable. Furthermore, wartime research did not free investigators of their professional obligations entirely. Richards noted that "although in most cases financial loss to the nonprofit institutions was avoided, there can be no question of the strain put on many of the investigators themselves."[35] This was es-

Figure 3.2. Federal Institutions Associated with World War II Vaccine Programs

BICIED	Board for the Investigation and Control of Influenza and other Epidemic Diseases, later renamed the Army Epidemiology Board
AMS	Army Medical School
CMR	Committee for Medical Research
CWS	Chemical Warfare Service
DMS	Division of Medical Sciences
FSA	Federal Security Agency
NAS	National Academy of Sciences
NIH	National Institute of Health
NRC	National Research Council
OSRD	Office of Scientific Research and Development
OSS	Office of Strategic Services
PHS	Preventive Health Service
PMD	Preventive Medicine Division
SGO	U.S. Army Surgeon General's Office
WBC	War Bureau of Consultants
WRS	War Research Service

pecially true in universities and medical schools, where, in addition to performing research under war contracts, research scientists and physicians were expected to train new doctors at an accelerated rate.

War placed high demands on the vaccine industry as well, pressuring developers to invest in an area with poor long-term commercial prospects. The introduction of sulfa drugs in the 1930s, and their success in combating bacterial infections, bred widespread pessimism about the future profitability of bacterial vaccines. Developers feared that antiviral medications would not be far behind. According to an internal memo at the National Drug Company, one of the oldest and most reliable vaccine producers for the military, "the attitude at the time was that chemotherapy would eventually bring about the complete dissolution of the biological industry."[36]

The Depression had further weakened consumer demand. By the late 1930s, many companies were reducing their investments, closing plants, and consolidating activities. Even the National Drug Company had embarked on a retrenchment program. They closed a plant in Germantown, Pennsylvania, and consolidated all activities into their remaining plant in Swiftwater, Pennsylvania.

As war became imminent, manufacturers reversed their retrenchment programs and scaled-up production to supply the armed forces. To meet these high-volume orders, producers agreed to expand manufacturing facilities, knowing that they would be left with excess capacity when demand slumped after the war. National Drug expanded its biological facilities to five times the original production capacity, to furnish the military with smallpox vaccine, tetanus, and gas gangrene antitoxin.

Demand was almost entirely determined by military orders. These orders, while large, were sporadic, and the government, as the largest single buyer of vaccines, could negotiate lower prices. These conditions did not translate into large profits for industry.

Looking back on these events, it seems remarkable that the military was able to procure nearly all of the vaccines it needed, including those with no foreseeable postwar market, through commercial channels. The military procured not only vaccines for influenza and tetanus, which did have a commercial future after the war, but also a large number of limited-use vaccines for plague, botulism, and Japanese encephalitis.[37]

It is tempting to conclude that patriotic concerns trumped traditional profit motives. While companies certainly took every opportunity to reinforce this interpretation, industrial decision making during national crises is more complex. Historians often reflect on the capacity of a crisis to inspire large-scale political and economic reform. Paul Adams argues that crises empower the government to act unilaterally and to effect change: "World War II," he contends, "constituted a crisis in which state and capital were vulnerable to an external military threat on one hand, and to internal pressures from below (due to a compelling need for working class labor, loyalty and sacrifice) on the other. In response to these threats, the state acted authoritatively and relatively autonomously vis-à-vis the capitalist class. Despite their suspicion of and hostility toward the state, capitalists were forced by the exigencies of war to submit to these statist, collectivist developments."[38]

Government-orchestrated vaccine development programs could be considered examples of such "statist" developments in action, but there is little overt evidence that the vaccine industry suffered from a sense of grudging submission. Company presidents and research directors appeared genuinely eager and proud to assist the war effort. Bush and Richards, upon their appointments as chairmen of the OSRD and the CMR respectively, received countless letters from pharmaceutical, chemical, and biological houses, sending congratulations and offers of assistance.

Randolph Major, director of research at Merck & Company, wrote Bush to assure him, "If we can so arrange our program as to be of help in the Defense Program we shall be glad to do this."[39] Major followed up with a brochure detailing the research and manufacturing capabilities of the company.[40] Similarly, Hans Molitor, director of the Merck Institute of Therapeutic Research wrote Richards: "[I will] offer you whatever help or assistance the Merck Institute or I personally might be able to give you. Of course you are familiar with our general research program, which is right in line with national defense, and with our facilities. However, since you were here last we have further grown, both in personnel and facilities."[41] Richards, prior to accepting his position on the CMR, had served as a scientific advisor to the Merck Institute since 1929, so relations between CMR and Merck in particular were friendly and familiar.

By all appearances, relations were amicable among the institutions the WRS contracted to perform BW research as well. According to WRS director George Merck, the "majority of institutions approached recognized the importance of the biological warfare program and were eager to participate in this war effort. They generously loaned the services of their highly trained staffs and made available, free of charge, their laboratories and equipment."[42]

Cordial—if not enthusiastic—relations existed in part because an external threat united their interests. An expectation of public relations benefits, coupled with the knowledge that all arrangements were temporary, may also account for industry's willingness to overlook immediate economic hardships. Industry may also have seen the war as an opportunity to boost flagging productivity.

The National Drug Company emphasized the connection between vaccines and national security for this purpose. On December 12, 1941, every employee received a memo entitled "Remember Pearl Harbor." It explained: "With the actual declaration of hostilities, this plant becomes, from a medical standpoint, a vital link in the chain of National Defense, therefore it behooves us to observe all precautions to maintain an uninterrupted flow of production."[43] Regardless of motive, rather than "submitting" to these programs, the vaccine industry appears to have seized upon the opportunity to improve public relations and boost productivity.[44]

Following a series of strikes in the defense industry, war planners also made explicit efforts to stoke patriotism and to equate productivity with freedom. To make industry employees feel like vital members of the war effort, the military awarded several commercial vaccine producers with the Army–Navy "E" award, for excellence in the production of war equipment. Addressing overworked production crews, Lieutenant Colonel R. R. Patch emphasized the national security value of the vaccines they were producing: "Some of you may have questioned whether you were doing your part in the war effort. Guns, tanks, planes and other agents of destruction are quickly recognized by everyone as war materials. But those things that save life or prevent the loss of life are not so easily recognized. Yet they may be just as important as the destructive implements. Those guns, tanks, and planes are useless without physically fit men to operate them. You men and women have played an important role in preventing sickness and keeping men fit to fight."[45]

The Navy celebrated technological innovation and high-volume industrial production as practical means to lofty ends. With the presentation of each "E" award, all employees were reminded that "by their unflagging spirit of patriotism . . . by their acceptance of high responsibility . . . by the skill, industry, and devotion they [were] showing on the production front of the greatest war in history . . . they [were] making an enduring contribution not only to the preservation of their country but to the immortality of human freedom itself."[46]

Equating industrial productivity with freedom also offered an attractive public relations opportunity. Drug stores displayed posters depicting a pharmacist dispensing medicine, interposed between a picture of the U.S. industrial complex and the shadow of a saluting soldier. Apart from stoking patriotism and boosting sales, this image allowed the pharmaceutical industry to present itself as the defender of the nation's health and security and it allowed corporate liberals to depict science and industry in support of the national welfare.

Federal programs administered through the OSRD and SGO leveraged this wartime spirit of cooperation to forge new interinstitutional and interdisciplinary alliances. While war provided the impetus for novel collaborative arrangements, wartime development programs ensured their success. The top-down administration of these programs enabled the rapid consolidation and application of preexisting knowledge. This arrangement was particularly effective for vaccine development, which requires the contribution of disciplines ranging from epidemiology, pathology, immunology, bacteriology, virology, and bioprocess engineering. Wartime development programs united individuals with this diverse range of expertise under a clear objective: to develop, test, scale up, and manufacture a prespecified set of vaccines. Under this integrated structure, the OSRD, SGO, and WRS accelerated development times by transferring people, technology, and ideas to the projects that needed them most.

Top-down administrative structures also ensured that vaccine research objectives matched military needs. The BICIED and CMR worked closely with military planning committees to align what was feasible with what was needed. Individual members of the CMR, for example, held joint membership in the Division of Preventive Medicine in the SGO, the Navy Bureau of Medicine and Surgery, the BICIED, the CWS, the Office of the Quartermaster General, and in the Office of the Air Surgeon. In this manner, the CMR ensured that military needs were well articulated and accounted for in all research and development planning sessions. The BICIED and CMR

Figure 3.3: The Pharmacist Courtesy of Merck Archives

also provided outside scientific and medical expertise through advisory committees and professional networks. Outside experts vetted development projects for scientific feasibility before they received military sponsorship.

Research and military objectives were so well coordinated, in fact, that some vaccines were developed in time for specific military missions. The botulinum toxoid, for example, was developed for D-day in response to inaccurate OSS reports that the Germans had loaded V-1 rockets with the toxin. Similarly, the Japanese encephalitis vaccine was developed in anticipation of an Allied land invasion of Japan.

In 1945, James Conant, director of the NDRC under Bush and president of Harvard University, gave a succinct summary of the method and rationale behind the OSRD's success. In a letter to the editor of the *New York Times* he wrote, "There is only one proven method of assisting the advancement of pure science—that of picking men of genius, backing them heavily, and leaving them to direct themselves. There is only one proven method of getting results in applied science—picking men of genius, backing them heavily, and keeping their aim on the target chosen. OSRD . . . had achieved its results by the second procedure which is applicable to government-financed research in war time because the targets can be chosen with a reasonable degree of certainty . . . its objective was not to advance science but to devise and improve instrumentalities of war."[47]

Once research and development priorities were set, scientific, technical, and operational decisions lay with the commission directors contracted through the BICIED and with the principal investigators under the CMR. This practice was consistent with a principle that Vannevar Bush referred to as "giving a man his head." He argued that "this is more than a matter of scientific freedom, important though that principle is . . . it is entirely possible to give a man his head and yet to specify by agreement with him his objectives."[48]

These programs also benefited from simple, direct reporting between the scientists and high-level SGO, BICIED, and CMR officials. In each case, vaccine development programs placed managerial authority with individuals who had the greatest scientific expertise.

This top-down management structure eliminated layers of bureaucracy and contributed to speed and efficiency.

As Conant's comments suggest, the success of World War II vaccine development programs was due less to scientific breakthroughs than to the ability of these programs to distill and apply existing knowledge. In many cases, the basic knowledge required to develop a new vaccine had been available since the 1930s or earlier. Barriers to the development of these vaccines were not scientific, but organizational, and they were best overcome through the coordination provided by these targeted research and development programs.

~ The pneumococcal capsular polysaccharide vaccine was one of the most radical innovations to come from World War II vaccine development programs. Not only did this vaccine prevent pneumococcal infections for the first time, but it used an entirely new method to confer immunity. As with most World War II vaccine development successes, a basic understanding of both the pathogen and the disease had been established before the United States entered the war. It was not until the war, however, that federal programs facilitated efforts to assemble this knowledge into a new vaccine application.

In 1927, Wolfgang Casper and Oscar Schieman at the Koch Institute in Germany, demonstrated that vaccines made from purified pneumococcal capsular polysaccharides would immunize mice against infection from the pneumococcal strains used to make the vaccine. This research drew from Michael Heidelberger's research at the Rockefeller Institute, which indicated that substances other than proteins could have antigenic properties. A capsular vaccine was preferable because, unlike a whole cell vaccine that could cause disease if bacterial cells are improperly killed or weakened, polysaccharide capsules could generate immunity without the risk of causing disease.

In the 1930s, Thomas Francis, Walter Tillet, and Lloyd Felton conducted a series of laboratory and clinical studies at the Rockefeller Institute to demonstrate that pneumococcal capsular polysaccharides were immunogenic in humans. Their research identified the particular antigens on polysaccharide capsules responsible for inducing an immune response and determined how to isolate and purify these antigens to produce a vaccine. Pilot vaccines were tested during the 1930s on volunteers from the West Coast Civilian Conservation Corps.[49] These studies provided early evidence that polysaccharide vaccines offered sufficient levels of safety and efficacy.

Industry and academia lost interest in pneumococcal vaccines after the antibiotic sulfapyridine was introduced in 1939, and physicians began substituting pneumococcal antisera with sulfonamides. Sulfonamides were less expensive, easier to administer, and considered safer and more effective against a wide spectrum of pneumococci.

All efforts to develop a new vaccine to induce active immunity would likely have come to a halt if the military did not have an enduring interest in population-based preventive measures. Vaccines remained more attractive to the military for the simple reason that a vaccine would reduce the overall number of sick days for the armed forces more effectively than therapeutic measures. To this end, the BICIED formed the Commission on Pneumococcal Diseases to continue the search for a pneumococcal vaccine.

Since the scientific feasibility of this vaccine had already been established, the remaining challenge for the pneumococcal commission was to apply this knowledge. Their job consisted of coordinating the expertise of scientists, engineers, epidemiologists, and physicians to identify the appropriate bacterial subspecies, or serotypes and then to develop, scale up, and test a vaccine containing these serotypes.

The commission conducted a survey of the most prevalent pneumococcal types in the military. Based on this information, Michael Heidelberger, who had joined the commission, supervised the production of purified polysaccharides types 1, 2, 5, and 7 at a pharmaceutical plant of E. R. Squibb & Sons. As large lots of purified polysaccharides became available, Heidelberger performed a series of small-scale clinical studies to determine the optimal dosage of polysaccharide required to obtain and sustain a sufficient immune response.

Once Heidelberger established consistent production standards and optimal dosages, the BICIED was ready to test the efficacy of this new vaccine more broadly. The crowning achievement of wartime clinical research occurred under the auspices of the BICIED when Colin MacLeod, a captain in the U.S. Army Medical Corps, performed the first randomized, placebo-controlled trial of a quadravalent pneumococcal vaccine for the military.[50] This study enrolled over seventeen thousand men at the Army Air Force Technical School between 1944 and 1945. Half received the vaccine and half received a placebo. At the end of a seven-month observation period, four men in the experimental group contracted pneumonia, whereas twenty-six men in the control group contracted the disease. Heidelberger recalled that "the entire study, so beautifully organized and monitored under MacLeod's direction, showed that epidemics of pneumococcal pneumonia in closed populations could be terminated within two weeks after vaccination with the polysaccharides of the causative types."[51] This study, according to Heidelberger, set the standard for all future clinical trials.[52]

Much of the scientific groundwork for an influenza vaccine had also been established before the United States entered World War II. In 1933, Patrick Laidlaw, at the National Institute for Medical Research in London, isolated a filterable virus from a patient with influenza and determined that this agent (which became known as influenza type A) produced flu-like symptoms in ferrets. In 1940, Thomas Francis, at the New York University College of Medicine, isolated the type B strain of influenza. In the early 1940s, in Australia, Macfarlane Burnet developed methods for growing the virus in developing chick embryos. By the time the SGO formed the influenza commission in 1941, investigators had already established the etiology of the disease and developed methods for isolating, cultivating, and purifying components of an influenza vaccine. All that remained for the influenza commission was to determine methods to scale up the vaccine for industrial production and to evaluate it for safety and efficacy before administering it to the armed forces.

Francis, recently appointed director of the influenza commission, worked with contracted CMR scientists and industry to develop a vaccine. The CMR issued contracts to investigate technical aspects of influenza vaccine development. Under these contracts, investigators determined conditions for improving virus yields from embryonated eggs and methods for improving titration accuracy and purification. In particular, investigators determined that they could concentrate and purify portions of the flu virus more efficiently with a Sharples centrifuge than they could with traditional small-batch methods (red cell elution and precipitation).[53]

By 1945, the Army accepted the CMR's production methods as an alternative to these traditional methods. Months later, industry adopted CMR production guidelines as well, and expanded production to meet civilian markets by early 1946. The CMR's Sharples centrifuge purification techniques were considered state of the art in the industry until the 1960s, when National Drug adopted a new method (zonal centrifugation) from military labs.

As new methods of vaccine production became available, Francis worked with industry to improve the potency and purity of vaccine lots for clinical trials. The BICIED consulted with Parke Davis, Lilly, Lederle, Sharp & Dohme, and E. R. Squibb & Sons to produce sample lots of the vaccine. Francis would then test these lots in his own laboratory at the University of Michigan and compare notes with industry to develop uniform standards for potency and purity.

Together with the CMR, industry, and the SGO, Francis and his team standardized the production procedures, record systems, and viral and serological tests that would permit the uniform clinical study of more than 12,500 members of Army Specialized Training Program Units (ASTP) across the country. These men were vaccinated in October and November of 1943, just weeks before an influenza epidemic hit the nation. Under the auspices of the BICIED, Francis conducted field studies with the influenza A vaccine.

By early January, the influenza commission produced the first conclusive evidence that they had a safe and effective vaccine against epidemic influenza A. Based on these findings, the SGO requested ten million doses of the vaccine for the U.S. Army in case of another outbreak.[54] Thus, within two years of initiating a development program, the first influenza vaccine became available for general administration in the military.

In 1945, the BICIED had an opportunity to test the B-strain vaccine as well when a wave of influenza B passed through U.S. Naval and ASTP units in November and December. Since these units were already under observation at the University of Michigan and Yale University, the BICIED was able to compare hospital admission rates between vaccinated and unvaccinated groups to demonstrate the efficacy of this vaccine.

Not all vaccine development programs were equally successful. When the underlying scientific principles were well established, as was the case with pneumococcal and influenza vaccines, federal programs were able to develop new vaccines quickly. These programs faltered, however, when the underlying science was less well understood. As anthrax investigators discovered, it was neither possible nor desirable to exercise top-down control before a basic understanding of the disease was established.

The WRS contracted Louis Julianelle at the Public Health Research Institute in New York City to investigate the bacteriological and immunological aspects of anthrax. Julianelle found himself challenging pre-existing knowledge of the disease more quickly than he could build from it. He began by testing the effectiveness of commercial antisera, only to conclude that these antisera could not mitigate anthrax infections. He determined that sulfa drugs were useless as well. Penicillin, however, was effective in high doses. As penicillin was in short supply, he then attempted, without success, to develop vegetative, capsular, and spore vaccines.

Julianelle reported to the WRS: "Text-book accounts all indicated that the subject was pretty well closed as far as immunization, and specific serum therapy was concerned. It came a good deal as a surprise to discover how incomplete or unreliable was indeed the existing knowledge. Consequently, it has been necessary to rework a number of the more fundamental phases of 'N' [code for anthrax] with the result that progress has been slow and the lag period of relearning prolonged."[55]

Julianelle's experience was characteristic of many WRS projects. While the WRS mimicked the OSRD's administrative structure and targeted research strategies, it had a less impressive record in vaccine development. Botulinum toxoid was the only human vaccine to come out of the WRS program. This may have been due in part to the fact that WRS investigators were often asked to work with more exotic organisms, such as coccidioides, brucellosis and tularemia, which had not been extensively studied by the general scientific community.

Without a fundamental understanding of many of the biological warfare diseases under study, the need for basic scientific research undercut targeted research and development objectives and demoralized WRS participants. Some members of the WRS advisory committee (known as the ABC committee) began to feel the WRS program lacked direction.[56] When the ABC committee was reconstituted in 1944 to oversee the expansion of the biological warfare program within the CWS, Perry Pepper, a physician at University of Pennsylvania Hospital and director of the new committee (known as DEF) had a difficult time encouraging colleagues to renew their commitment.[57]

Ernest Goodpasture, a professor of pathology and viruses at the Vanderbilt University School of Medicine and a former member of the ABC committee, declined Pepper's invitation to join its replacement, the DEF committee. Goodpasture explained that he had for some time tried, without success, to discern the objective of the ABC committee: "I have the feeling that no critical survey of existing potentialities in relation to military needs has been undertaken or at least has resulted in the definition of objectives suitable to guide the work in laboratories. It appears that the military interests are relying entirely on the laboratory worker to suggest application to military uses concerning which the laboratory investigator has no knowledge himself or direction from military experts."[58]

A murky understanding of many of the pathogens of concern made it difficult for scientists and military experts to identify or defend clear development objectives. Scientific understanding was in its infancy, and no amount of money or appropriate organizational support was likely to yield new vaccines at that time.

~ WORLD WAR II VACCINE development programs were most successful when they drew on the military's direct experience with a disease, as was the case with pneumococcal pneumonia and influenza. Success can be attributed in part to the military's status as a "lead user," a term coined by Eric von Hippel. Lead users "face needs that will be general in a marketplace but they face them months or years before the bulk of that marketplace encounters them, and lead users are positioned to benefit significantly by obtaining a solution to those needs."[59] "Users," according to von Hippel, "are the generators of information regarding their needs."[60] The military, for example, has the most information about the particular medical interventions they need and the ways in which they need those interventions to work. "Manufacturers, in contrast, have poorer information on users' needs and use contexts, and will prefer to manufacture innovation for larger, more certain markets."[61] User needs, therefore, are a form of sticky information in that they are hard to know outside of close relationship with the users themselves. In short, lead-user collaboration "helps to reduce information asymmetries between users and manufacturers and so increases the efficiency of the innovation process."[62]

One serendipitous outcome of World War II development programs was that industry was able to take full advantage of the military's experience with disease control and vaccine development. The military had, and continues to have, vaccine requirements that exceed those of the general population. Not only does the military have to defend itself against biological weapons, but the effects of natural disease are exaggerated under military conditions. Training camps and battlefields produce dense populations of physically exhausted and emotionally stressed individuals, which allow communicable diseases to spread like wildfire. Consequently, respiratory and common childhood diseases, such as measles, mumps, and rubella, present a greater threat to adults in military contexts than in civilian life.

Troops are also exposed to a wider range of diseases. In the field, troops are subject to unfamiliar disease vectors, such as mosquitoes that transmit yellow fever, malaria, and dengue, or to arthropods that spread rickettsial diseases. A memo to the U.S. Army R&D Command explained that "because of the requirement placed upon the Army to fight anywhere in the world at any time, certain other areas of the world must be carefully scrutinized [for disease threats that would compromise military campaigns]. The Tropics, including South Asia, the Middle East, almost all of Africa, most of the Latin American countries, the majority of the Pacific Islands, and a portion of Australia are of particular interest . . . Diseases encountered include many to which the American soldier has not previously been exposed. To a force operating in these areas on widely dispersed battlefields, this fact would mean hospitalization of roughly half the force within a year."[63] In the Korean War, for example, "from July 1950 through July 1953, admission to U.S. hospitals for disease was 66 percent of all admissions."[64]

Before the Gates Foundation began supporting public-private product development partnerships at the turn of the twenty-first century, the U.S. military had been one of the few institutions with the skills, resources, and motivation to organize effective research and development programs for diseases in the developing world. The reasons for this are not intentionally humanitarian. In contrast to private medicine, in which care is triggered by illness, military medicine is focused on improving daily noneffectiveness rates. The military has a greater incentive to develop population-based preventive measures, such as vaccines, which maintain mission readiness by reducing sick days and recovery time. Private medicine, by contrast, fosters a higher demand for therapeutic measures.[65]

Apart from a predisposition to devise preventive measures, scientists and physicians working in military contexts also had a number of specialized skills and organizational advantages. While the OSRD and SGO provided clear direction on research objectives and development needs, these scientists and physicians were also skilled at identifying new diseases in military populations and determining their etiology. Treating demographically homogenous populations that ate, slept, and worked under shared environmental conditions also made it easier to identify population-level characteristics of new or familiar diseases in military contexts than in civilian settings.

Military installations, with their advanced record-keeping systems, controlled populations, and high disease rates, also offered an unparalleled proving ground for new vaccines. One member of the Preventive Medicine Division noted that "the practitioner of military preventive medicine has at his disposal information on morbidity of a quality not available in any other social organization."[66] Francis observed that field trials at military installations afforded an ideal opportunity to assess the early vaccines developed by the Influenza Commission. The installations "were stable populations and subject to constant, uniform observations. It was possible to obtain participation of entire units so that vaccinated persons and controls could be properly designated rather than depending upon the less desirable and unpredictable use of volunteers."[67]

Even when well-planned clinical trials did not occur, the widespread use of a vaccine in military populations could offer *de facto* evidence of safety and efficacy. For example, the military experienced such high rates of safety and presumed efficacy with their tetanus vaccine that the American Pediatric Association recommended routine use of this vaccine in the general population in 1944, without formal clinical trials.

As a lead user, the military often revealed problems and guided improvements before particular vaccines were administered to the general population. Prewar versions of the tetanus vaccine, for example, were highly reactogenic, but in-house research conducted by the Preventive Medicine Division identified peptones (protein derivatives) in the nutrient media used to grow the bacteria as the cause of allergic reactions. The SGO contracted J. Howard Mueller at Harvard Medical School to develop a synthetic alternative to these vaccine broths, and industry adopted the new process formula for all future contract orders.

Military use of the yellow fever vaccine at the beginning of the war revealed critical flaws as well. Scientists at the Rockefeller Institute added human serum to their vaccines to prevent them from deteriorating and losing effectiveness. Soon after the army administered this vaccine, which had not been tested for safety, troops began to contract "serum sickness." It was later determined that blood contaminated with hepatitis B had been mixed into the vaccine, resulting in fifty thousand cases of jaundice and sixty-two deaths in 1942. The PHS developed a safer, non-serum-based version of the vaccine that same year. As a direct result of the military's experience, scientists abandoned the practice of using human serum to stabilize vaccines.

While close collaboration with the military produced a record number of new and improved vaccines in a short time frame, this partnership had mixed outcomes for industry. On the one hand, many of the companies that worked closely with the OSRD and the BICIED found themselves at an advantage after the war since they had been forced to adopt new production methods. According to a National Drug memo, in order to meet military contracts, "it was necessary also to forsake academic methods of production and to adopt more of the mass production methods found in other industries." The memo observed, "We can well assume that our competitors were faced with many of the same circumstances and that they too have adopted the same principles of operation. Subsequent visits to their laboratories have substantiated this assumption."[68] Companies that invested in these industrial techniques were well positioned to capitalize on growth in vaccine markets in the postwar era.

On the other hand, military needs do not always reflect civilian market demands, as evidenced by the military's interest in tropical diseases and biological warfare. Even when military and civilian interests coincide, as they do for many respiratory diseases common to North America, close collaboration with the military has on occasion put industry too far ahead of civilian market needs, making it difficult to reap financial rewards.

For example, after sulfonamides were introduced, both industry and academia lost interest in a pneumoccocal vaccine. Only the military continued to invest in a research program. By 1944, the decision to continue pneumococcal research proved wise, as training bases began to reveal holes in the new antibiotic armamentarium. An army air force training base in Sioux Falls, South Dakota, began to experience recurrent epidemics of pneumococcal pneumonia despite the use of sulfa drugs. The army urged E. R. Squibb & Sons to prepare a quadravalent vaccine for clinical trials based on MacLeod's work with the commission. This was a risky proposition for Squibb since it required them to build entirely new production facilities for an unproven vaccine.

The army medical branch also asked Heidelberger to continue

work on another capsular polysaccharide vaccine with pneumococci types 1, 2, 5, and 7.[69] Subsequent studies by Heidelberger demonstrated the safety and efficacy of hexavalent vaccines as well.[70] On the basis of Heidelberger and MacLeod's data, the army urged Squibb once again to develop and market two different hexavalent pneumococcal capsular polysaccharide vaccines. By 1945, Squibb had poured millions into building and staffing new plants, simply to produce the vaccine for clinical trials. In 1948, they agreed to market one vaccine for adults and one vaccine for children, each containing a slightly different array of serotypes.

Despite their proven safety and efficacy, pneumococcal vaccines were a commercial failure. Antibiotic resistance to pneumococci was not as widespread in civilian populations as it was in the military by the end of World War II. Doctors, convinced of the value of antibiotics in treating pneumococcal infections, preferred therapeutic to preventive measures, as a general rule.[71] As the use of antibiotics grew, Squibb could no longer afford to stay in the pneumococcal vaccine business. In 1954, the company terminated production of the vaccine. Levels of antibiotic resistance in civilian populations eventually caught up with those of military populations, and commercial interest in pneumococcal vaccines resumed by the 1970s, but it was far too late for Squibb to recover from its financial losses.

The military has been a highly sensitive indicator of future disease threats, but it has put industry on the "bleeding edge" (rather than the leading edge) of vaccine development on more than one occasion. History repeated itself in the 1970s when the military encouraged industry to develop meningococcal vaccines—well before commercial markets could support industry participation. Industry eventually had to terminate production of this vaccine as well. As was the case with pneumococcal vaccines, the decision to terminate meningococcal vaccines proved premature, since civilian vaccine protection needs eventually caught up with those of the military.

~ WHILE WARTIME PROGRAMS had ambiguous outcomes for industry, their long-term impact on society has been more dubious still. World War II programs developed an unprecedented number of new or improved vaccines, but many scientists working under government biodefense contracts made more progress in learning how to spread disease than in learning how to prevent it. Reflecting on the history of war and technology, William McNeill observed that there was "moral ambivalence implicit in every increase in human power to manage and control our natural and social environment."[72] Though he did not consider the moral ambivalence of World War II vaccine development programs, there were few projects that embodied a more contradictory mix of national welfare and warfare objectives.[73]

Wartime advances in methods for growing high-yield plague and typhoid cultures, for example, could be used either to improve vaccine production or to accelerate the development of biological weapons. The federal government often contracted scientists to ensure that their inventions were available for *both* purposes. Rene Dubos, a microbiologist, and Henry Hoberman, an infectious disease specialist, led two projects at Harvard Medical School during the war.[74] The public was aware only of the first project, which attempted to improve typhoid vaccines. The second project, code named project "Y," determined methods for the mass production of the Shiga bacillus. As was often the case, the offensive objectives proved easier to accomplish than the defensive ones. While Dubos's group made no significant progress in the development of a more effective vaccine, it succeeded in developing production processes that produced high yields of the bacilli in record time.[75] With several other scientists working to obtain high yields of bacterial cultures for offensive and defensive purposes, the OSRD rapidly transferred Dubos's high-yield production process to other scientists under contract.

Dubos and Hoberman were not the only scientists to accept CMR and WRS contracts to improve the technologies of disease prevention and creation simultaneously. Karl Meyer, a biochemist at the George William Hooper Foundation in San Francisco, accepted contracts through both organizations to develop vaccines and weapons with plague-causing bacteria. Norman Topping, a research scientist at the NIH, accepted contracts to conduct offensive and defensive research on typhus. Lee Foshay, a professor in the Department of Bacteriology at the University of Cincinnati, and Cora Downs, a professor in the Department of Bacteriology at the University of Kansas, each worked independently on techniques to both prevent and induce tularemia. Similarly, Forest Huddelson, from the Department of Bacteriology at Michigan State College, worked on brucellosis weapons and vaccines. J. J. Griffiths, from the NIH, also developed cholera-based weapons and vaccines.

The line between defensive and offensive research could be exceedingly thin and moral dilemmas were unavoidable. To test vaccines, scientists had to produce infectious agents so that they could perform challenge studies. To handle biowarfare agents responsibly, scientists often had to develop vaccines. It made sense to accept contracts that examined both sides of this problem.

Though it is hard to know how each individual resolved the moral dilemmas of offensive research, Meyer justified his work in part by pronouncing that he would not proceed with offensive research until he developed effective defensive measures. "Irrespective of what we may ultimately be able to accomplish in the offensive direction," he explained, "I am more than ever convinced that we must first and foremost plan to develop the defensive. The risk to workers is too great to venture even a pilot experiment on a small scale."[76]

Meyer did, in fact, succeed in developing an effective plague vaccine before he moved on to study methods of weaponization. He proved to be an exception. According to activity reports from the WRS advisory committee, few scientists managed to develop effective vaccines under WRS contracts because, in many cases, the diseases under study were poorly understood.[77] Many other scientists may have shared Meyer's aspirations, but ultimately their efforts to mass-produce and devise methods for the dissemination of various pathogens were more successful.[78]

It is likely that some participants were impatient with ethical delicacies in a time of war. At the very least, it was socially acceptable to approach the subject with irreverent humor. Several years later, in response to a request to join Merck's board of directors, Bush launched into a riff against medical ethics:

> I don't suppose we would get into a lot of medical ethics; I'm sure I couldn't take that. I've had measles and flu and lot of things that ordinary people have, but I've never had ethics, and I know medical ethics doubles people all up sometimes. You have to look out for these places where they get all hot and bothered about curing folks' diseases for them, there are likely to be a lot of wild germs around that they are practicing on, and one of them might get in your mucous membrane or your tibia or someplace where germs multiply and give you ethics. I've heard that when people get a real dose of medical ethics they actually go nuts; I don't mean they bite the furniture or assault the keepers, but they get irrational and their logic becomes screwy. I'm not too logical anyway when it comes to theories of human relations and highfalutin social schemes, and I'm sure I couldn't stand an attack of medical ethics. If you think there's any danger of infection we ought to quit right here.[79]

George Merck, director of the WRS, and later a special consultant on biological warfare to the secretary of war, was acutely aware of the moral ambivalence inherent in biodefense research. Following the war, however, when he returned to his duties as president of the pharmaceutical company that still bears his name, he emphasized the positive spinoffs of offensive biological research activities. In a 1946 article regarding U.S. biowarfare activities, he asserted:

> There cannot help but be important advances in knowledge—many of them fundamental—and gains in scientific achievement—many of them capable of practical application. In fact, it is quite impossible for work to be done in this field without such results. It is inherent in the nature of the work. Perhaps no other type of warfare can bring with it such a guarantee of good: economic advantages in agriculture, parallel gains in animal husbandry, and, above all, vital contributions to the fight against human ills and suffering . . . While we perfect a

> biological weapon, we perfect the defense against it, thereby destroying the weapon. Would that all weapons of war could be liquidated from the earth as simply as this.[80]

Better welfare through warfare? In light of the final WRS activity reports, which outlined a far greater number of offensive than defensive developments, Merck's arguments appear disingenuous. They do, however, shed light on the peculiar rationale that guided scientists through offensive biological research and development activities.[81]

Merck's comments also contain a kernel of truth. Heightened concern for biological threats during this period inspired unprecedented military investments in vaccine research and commercial vaccine development with positive outcomes for public health. These vaccines were often crude by today's standards, and sometimes unsafe, but on the whole they saved lives, maintained the operational readiness of the armed forces, and paved the way for future improvements. World War II development programs produced the first licensed vaccines for influenza, pneumococcal pneumonia, and plague. They also developed a new typhus vaccine after establishing that the former vaccine had lost potency. In conjunction with the WRS, they developed the first botulinum toxoid and Japanese encephalitis vaccine. They also made significant improvements to existing yellow fever, cholera, smallpox, and tetanus vaccines. Improvements to the safety of the smallpox and tetanus vaccines, in particular, facilitated wider use of these vaccines in the general population after the war, making a significant contribution to public health.

Why were vaccine development efforts so much more productive during this period than any other period in the twentieth century? In part, a sense of national urgency to defend against war-enhanced disease threats fostered unprecedented levels of federal support and a spirit of cooperation between military, industrial, and academic scientists. Additionally, targeted research and development programs, administered under the federal government, provided a structure that productively channeled this spirit of cooperation. A third, and more underappreciated source of innovation, however, derived from the participation of the military itself. World War II development programs paired industrial vaccine developers with a key lead user of vaccines to develop an extraordinary number of new or improved vaccines.

4
Wartime Legacies

THE BIRTH AND EXPANSION of research and development–based firms after World War II is a common theme in twentieth-century U.S. industrial history.[1] Just as wartime research in electronics, aircraft design, and jet propulsion provided a foundation for the expansion of postwar industries, it seems axiomatic that vaccine research would have transformed what was formerly a cottage industry into a full-scale research and development enterprise. In the immediate aftermath of World War II, however, the growth of the U.S. vaccine industry was far from assured.

Prior to the war, U.S. manufacturers would often look to their European counterparts for new discoveries and applications. In a letter to the director of Abbott Laboratories in Chicago, Richard Slee, director of the vaccine laboratory that would become the National Drug Company in Swiftwater, Pennsylvania, acknowledged, "America may be a very bright nation, but between you and I, we are really nothing but a nation of assemblers and we have built up our reputation most largely on adopting European ideas, buying their stock, stamping it together and putting a nickel plate or polish on it and calling it a product of America."[2]

The war cut supply lines to European research, equipment, and production, forcing U.S. manufacturers to develop their own capabilities. While firm capabilities had improved, military demand for vaccines fell after the war, leaving companies with excess production capacity and huge inventories of vaccines with little to no domestic market. Sulfonamides and the recent introduction of penicillin nearly eliminated demand for bacterial vaccines. The influenza vaccine, which was expected to have wide market appeal, was not doing well commercially because it caused sore arms and fevers. Wartime testing had also demonstrated that several widely sold antitoxins, toxoids, and vaccines in use before the war were ineffective and many were taken off the market, which eliminated another traditional source of revenue for industry.

Federal vaccine development programs disbanded after the war, depriving industry of an important source of innovation. The Committee on Medical Research (CMR) was dissolved and the War Research Service (WRS) became part of the Chemical Warfare Service (CWS), which was renamed the Army Chemical Corps. Only the Board for the Investigation and Control of Influenza and other Epidemic Diseases (BICIED), renamed the Army Epidemiology Board (AEB) in 1946, retained a research arm.

With the termination of military contracts, commercial biological houses began to reconsider their prewar retrenchment programs. According to one internal memo at the National Drug Company, "The major portion of government contract work was completed by the end of 1946. At that time the attitude of management . . . was somewhat pessimistic regarding the volume of biological business to be anticipated during the post-war era. Plans were made to consolidate facilities and some attempts to increase efficiency were made."[3]

Meanwhile, the few government labs that had assumed manufacturing responsibilities during the war were anxious to drop these duties in order to reclaim lab space and time for research. The secretary of war directed the chief of the CWS to shift all military vaccine production into the private sector "to the maximum practicable extent."[4]

Ralph Parker, director of the PHS Rocky Mountain Laboratory, questioned the role of the Public Health Service in vaccine production as well. His lab had produced an aqueous–base yellow fever vaccine, a typhus vaccine, and a Rocky Mountain spotted fever vaccine for the military. Now, however, he urged the director of the National Institutes of Health (NIH) to shift all government manufacturing responsibilities to industry. This, he cautioned, would not be easy: "I have been talking recently with Harold Cox of Lederle Laboratories . . . Certainly there is little actual desire on the part of the Lederle Laboratories to take over, due apparently to the fact that there would be little in it for them financially."[5]

Remarkably, however, government efforts to shift the manufacturing burden often succeeded. In September 1948, Parker turned responsibility for the Rocky Mountain spotted fever vaccine to Lederle and Sharp & Dohme. In the early 1950s, the Public Health Service (PHS) also persuaded National Drug to manufacture the yellow fever vaccine.

Industry did have a few reasons for cautious optimism. A sudden jump in birth rates following the war boded well for pediatric markets. Widespread vaccine use in the armed forces had also reduced public fears of vaccination. As servicemen returned home—most of whom had received multiple immunizations with no ill effects—vaccines began to develop a reputation as harmless, routine interventions.[6] "War," one company memo noted, "was the greatest test of the efficacy of biologics," and it spurred cultural acceptance of vaccines among doctors, scientists, and the public.[7] Thanks to wartime immunization programs, "12 million men returned to their homes with at least some knowledge of how and why they are used."[8] Public education campaigns further encouraged acceptance. The memo noted that "the American public is becoming better informed through current periodicals, movies, and broadcasts about the 'shots' that they formerly dreaded and now more than often request."[9]

In a few short years, the Korean War boosted military demand for vaccines once again. Additional orders did not, however, always improve the bottom line. According to National Drug Company records, government contracts connected with the Korean War pushed sales to an all-time high of $2 million in 1951.[10] Yet net earnings continued to drop, as National struggled to balance development costs against low returns on government contracts.

The federal government began to invest in biomedical research during this period as well. Some of these investments were ideologically driven. Bush's treatise—*Science: The Endless Frontier*—reinforced the idea that a strong scientific base was an investment in the economic health and security of the nation. Wartime research contributions, he argued, provided ample political justification for continued federal investments in science and technology.

Bush's belief in the social value of science mirrored a broader "postwar consensus" that federal investments in military research and development, in particular, were also investments in the economy and the human condition.[11] Two colonels from the Army Medical School detailed this concept in a 1951 article for the *New England Journal of Medicine:* "In accordance with our democratic traditions, a small nucleus of Regular Army specialists has combined with the civil medical profession to forge vast and effective organizations in time of war. In peace, an increased sharing of responsibility has eliminated barriers between military and civilian medicine and has fostered the fundamental concept that government agencies represent the will of the people. Proof of the validity of these principles lies in the steady reduction in disease morbidity and fatality of the injured."[12]

A collection of well-placed science advocates fueled federal support for biomedical research throughout the postwar era. Alfred Newton Richards, former chairman of the CMR and president of the National Academy of Sciences, argued that "the experience of OSRD has proved that none of the universities which were called upon for OSRD work could afford to undertake it on the scale which the emergency demanded at the expense of its own resources. Hence, if the concerted efforts of medical investigators which have yielded so much of value during the war are to be continued on any comparable scale during the peace, the conclusion is inescapable that they must be supported by government."[13] James Simmons, chief of the Preventive Medicine Service, argued for continued federal sponsorship as well, contending that "the security of the nation depends on the health and physical strength of all its people, both military and civilian, and that a continuing program of research in military medicine is essential to its security."[14] In an appeal to protect budgets in peacetime, he went on to explain, "The need for medical research by and for the Army bears no direct relationship to the size of the Army. The medical problems of a future war will be the same regardless of the size of the Army during the intervening years."[15]

Richards's and Simmons's arguments were well received. There was broad agreement among the public and within Congress that federal investments in science and technology improved the security of the nation, the health of the economy, and the human condition.[16] Agreement broke down when the discussion turned to questions of where and how the government should support biomedical research.

Congress debated proposals for several years following the war.[17] In 1950, when Congress finally agreed on a successor for the OSRD in the shape of the National Science Foundation (NSF), it was clear that the NSF would support only basic research. Responsibility for more applied forms of biomedical research would fall partly to military research institutions and partly to the NIH.

This was not the arrangement that Bush or Richards had envisioned. They favored an independent agency (which they called the National Foundation for Medical Research) to disperse grants-in-aid to medical schools and universities. In their view, no preexisting agency, including the NIH, was "sufficiently free of specialization of interest to warrant assigning to it the sponsorship of a program so broad and so intimately related to civilian institutions."[18] Bush and Richards wanted to turn the government's attention back to the universities. They were concerned that federal planners might grant too much credit for wartime innovations to the military labs themselves. Their proposed foundation was designed to expand on the CMR's success leveraging civilian science. Richards wrote: "It must be emphasized that there is little in war medicine that did not have its roots in civilian studies and practice. The pressure of war served chiefly to accelerate the development and large scale application of discoveries particularly applicable to military needs."[19]

When Bush's and Richards's agency failed to gain support, federal dollars for vaccine research began to flow to preexisting government research organizations, such as the NIH and, most notably, the Army Medical Graduate School (AMS) and the Army Research and Development Board. The AMS expanded their facilities throughout the 1950s and renamed the research branch of the AMS the Walter Reed Army Institute of Research (WRAIR) in 1955. By 1958, the Army Research and Development Board was reorganized into the U.S. Army Research and Development Command, which controlled a network of military labs and installations performing medical research. Eventually, the network expanded to include fourteen laboratories in the United States and overseas. After hovering between $10.1 million and $10.5 million for the first half of the 1950s, the command enjoyed successive budget increases after 1956, doubling their budget by 1961.[20]

In a postwar political climate that favored federal investments in military-sponsored biomedical research, and in the absence of a civilian successor agency to the CMR, military research laboratories gained new importance as national centers for vaccine development. By the 1960s and 1970s, the trend that Richards had observed during the war began to reverse itself and it began to appear that there was little in civilian vaccine development that did not have roots in military research programs. Military research made significant contributions to eight out of fourteen new vaccines licensed in the second half of the twentieth century: the adenovirus, anthrax, hepatitis A, hepatitis B, Japanese encephalitis, measles, meningococcal meningitis, and rubella vaccines.[21] Military sponsored research also contributed to incremental improvements to influenza vaccines, combined diphtheria tetanus vaccines, cholera, smallpox, and typhoid vaccines.[22]

Political and military interests influenced both the direction of vaccine research and the rate of innovation. The military's presence in Africa, Southeast Asia, and East Asia directed resources toward tropical diseases, such as malaria and dengue fever that were of little commercial interest to U.S. pharmaceutical firms.

WRAIR and the Armed Forces Epidemiology Board (AFEB) enhanced their epidemiological and clinical competencies through a network of international laboratories and field stations.[23] This infrastructure facilitated efforts to improve the Japanese encephalitis, typhoid, and cholera vaccines by working directly with the populations most affected.[24]

As the Cold War intensified, nuclear and biological warfare fears also fueled investments in military medical research and development. The Surgeon General's Office (SGO) wanted to develop and stockpile vaccines to ensure operational effectiveness and to stabilize populations in the wake of a large-scale nuclear or biological attack. One memo observed: "If we are to fight the kind of ground war that is projected for the future, or if we are to pull ourselves together after a massive nuclear attack and prepare to support any kind of war, we will be faced with circumstances under which our control of the environment will be in many cases very greatly reduced. Therefore the threat of infectious disease, such as I have noted above, may be vastly greater under these projected circumstances than it would appear to be if one merely looked around the world today."[25]

The joint chiefs of staff and the secretary of defense also wanted to expand military biological warfare research. The military had long maintained a "no first use" policy toward biological weapons, but the National Security Council (NSC) removed this restriction in 1956.[26] At a 1959 AFEB meeting, General Hays (surgeon general of the army) observed: "There has been in the last two years with the Army Medical Service rather a shift-over in the picture of BW [biological warfare]. I can remember just a few years ago when most of our people took the attitude that nothing had been proved in this field, and therefore, there was nothing to it. I think that the opinion of our people has changed, and that we all feel that BW exists as a potential weapon, and that we must very actively pursue research in this field."[27]

There was a growing recognition that vaccines had strategic value, not merely as a form of defense but as a diplomatic tool. Modern medicine, in general, and vaccines, in particular, had become useful in the battle for the hearts and minds in contested regions. One SGO memo argued that a military medical presence overseas "establishes warm personal and professional ties with key individuals and the populations generally of countries in so-called 'underdeveloped' regions—greatly enhancing the prestige of the U.S. among nations of the world."[28] These programs were lauded for many of their indirect strategic benefits: "the good will that they create, the contacts that they establish, and the entrée that they provide for professional, diplomatic and other contacts within the areas concerned."[29]

In 1962, Captain Leonard Freidman wrote an essay on "American Medicine as a Military-Political Weapon" for the *Army R&D Command Annual Report,* in which he argued for sending public health teams to Vietnam. He cited the success of General MacArthur's "use of American medicine in producing good will toward American aims" in postwar Japan.[30] That program had vaccinated eighty million Japanese against smallpox, thirty-six million against tuberculosis, and thirty-four million against cholera. He postulated that the "failure of American military-political missions to stabilize China between 1942 and 1949 could be based partially on the inability, during World War II, to utilize medicine as an opening gambit in approaching turmoil-ridden China."[31] He argued that "training of medical sub-professional personnel—medics, nurses, laboratory technicians, and teachers drawn from the people of a national minority—[could] lead to further acceptance of Western ideas and ideals." He suggested that: "Subsequently, the minority group may be led to a wish to provide its own military contribution to the Central Government, as a response to a feeling of conciliation and concern on the part of the government, demonstrated through introduction of modern medicine and education."[32] Such insights inspired the Kennedy administration to launch the Peace Corps and motivated the Johnson administration to try to pass an international health and education act, which would have integrated international medical aid with diplomatic objectives.[33]

On occasion, U.S. and Soviet medical teams would compete for the opportunity to curry favor with vaccines. A 1961 polio epidemic in Kyushu, Japan incited one such skirmish. The surgeon general's early morning meeting minutes reported: "The Russians have offered the Sabin vaccine for inoculations to the Japanese and naturally, they feel they cannot refuse this offer from a political standpoint. A polio inoculation program has never been established in Okinawa because they have a natural program of immunization.

Okinawa is a U.S. occupied and administered territory and it is believed (in order to avert adverse publicity) that these children should be inoculated quietly and quickly although preventive medicine people realize it is unnecessary. However, the Russians would enjoy using this as a means for their propaganda."[34]

For the most part, however, U.S. and Soviet activities in international medicine were characterized by cooperation and had positive consequences for international public health. The first live oral polio vaccine was developed in close cooperation with the Soviet Union. This transcontinental scientific collaboration began in 1956 when Albert Sabin transferred his strains to Soviet virology institutes for large-scale development, manufacturing, and clinical testing.[35] Once these trials demonstrated that the vaccine was safe and effective, the United States and the Soviet Union continued to collaborate, under the auspices of the World Health Organization (WHO) and other international agencies, to administer the vaccine abroad. Under this program, the number of polio-infected countries fell from 125 in 1988 to four in 2010.[36] In 1967, through the WHO, the United States and the Soviet Union engaged in another collaboration, which eradicated smallpox from the globe. The last known case of smallpox occurred in 1977, and the WHO declared victory against the disease in 1981.[37]

As Cold War investments in vaccine research grew, WRAIR emerged as a center of excellence for infectious disease research and vaccine development. According to one report, "Prior to World War II, medical research was rarely a full time duty assignment for medical officers, and still more rarely a career. Much of the research was done on an individual basis and on individual initiative."[38] After the war, however, military research labs boasted state-of-the-art facilities and starting salaries that were competitive with university and industry laboratories.[39] WRAIR offered an attractive career option for young scientists graduating in the postwar period and it was able to attract the nation's top graduates. WRAIR's prestige grew as its alumni proceeded to top positions in industry, academia, and government.

One alumnus noted that, in addition to being an inherently attractive place to work, WRAIR enjoyed unique human resource advantages before the military discontinued the draft in 1973. WRAIR, he explained,

> was where all the bright young doctors graduating from medical school would go. This was a great way to evade service in a productive manner. They would get all the best men. If you were about to be drafted and a professor had someone with tremendous aptitude, it would have been a waste to send them out to an aid station—so they were referred to the Walter Reed research program. They would do research in the field and had a base at Walter Reed. They were so good that for a time, the heads of many Infectious Disease and Pediatrics departments had been at Walter Reed. They were an illustrious set of alumni.[40]

Through WRAIR, talented research scientists received unique hands-on interdisciplinary training. Whereas academic researchers were encouraged to focus on fundamental questions, military research scientists were trained to work through the practical implications of their findings and to facilitate industrial adoption of their research. A 1957 manual outlining WRAIR's research objectives instructed staff scientists "to develop the production processes required to translate laboratory scale results for large scale production by industry."[41] To this end, WRAIR offered training fellowships that sent their scientists to industrial labs to learn large-scale production techniques for biological products.[42]

WRAIR bred the type of scientists who fit Bush's ideal view of science integrators. Bush often warned against the dangers of overspecialization in disciplines and once proclaimed, "I should be inclined to establish a Nobel Prize for the integrator and interpreter of science who can, in these days, serve his fellows far more than the individual who merely adds one morsel to the growing, and often indigestible, pile of accumulated factual knowledge."[43]

At WRAIR, there was no formal division of labor or strict departmental specialization. A former WRAIR scientist explained that they were explicitly "trained to go to the field and come back to the laboratories."[44] Another member of WRAIR observed: "U.S. Army scientists, as opposed to university scientists, are particularly strong at bringing a vaccine through the whole development spectrum . . . They had expertise at each stage. The military has unique assets/needs that are not driven by a fiscal bottom line but by the need to generate a product."[45]

Industry recognized the value of this new breed and began to turn to WRAIR scientists for collaborative research partners and new hires. In 1957, for example, the SGO reported that the "Department of the Army received a letter from the Pfizer Drug Company asking for five retired doctors who might be interested in some interesting and lucrative positions with their company."[46] In the same year that Pfizer was plumbing the ranks of WRAIR, Merck & Company hired Maurice Hilleman to head their new Virus and Cell Biology Division. Hilleman was among WRAIR's celebrated alumni, working in the Department of Respiratory Diseases from 1948 to 1956. Referred to as the father of modern vaccines, Hilleman is credited with developing more vaccines, and saving more lives, than any other scientist.[47] The National Drug Company imported several WRAIR scientists into key positions as well. This migration of talent strengthened informal military-industrial networks and fostered a highly productive collaborative community for vaccine development in the postwar era.

POSTWAR INDUSTRIAL VACCINE development drew heavily on the personal friendships, ideologies, and scientific networks forged through World War II. The legacy of wartime research and development programs was particularly strong at Merck. George Merck believed that scientific research would play a key role in the political and economic future of America and—no less importantly—his company. By raising the profile of scientific research at his company, he was able to shed the pharmaceutical industry's prewar reputation as a collection of underhanded snake-oil salesmen.[48]

After the war, Merck began to recruit prominent scientists onto his board, a task made easier by his participation in wartime research and development programs. As director of the War Research

Service, Merck came into contact with the top research scientists and administrators of his day, and he was positioned to persuade many talented individuals to join the Merck board of directors. These included Vannevar Bush (former director of OSRD), Edward Reynolds (former brigadier general, SGO, in charge of medical supplies), and Alfred Newton Richards (former director of CMR). John T. Connor (former general counsel for the OSRD) joined the company as a direct employee and eventually became Merck's successor as CEO.

Although Merck had a longstanding belief in the value of scientific research in industry, the practice of inviting noted scientists onto the board was new. Prior to the war, board membership was largely restricted to bankers and lawyers. By recruiting a large fraction of the OSRD, Merck imported a new collection of ideological perspectives, political tactics, and research practices that had gained currency during the war. Merck's emphasis on a scientist-heavy governance structure was, in itself, a wartime tactic that leveraged the authority and the apolitical appearance of scientific opinion.[49]

Richards was the first of an elite group of scientists elected to the board, joining Merck's directorship in 1948. Soon after electing Richards, Merck extended an offer to Bush as well. Merck conveyed his offer through Richards, who had worked closely with Bush on medical research at the OSRD. With a comical nod to the industry's prewar reputation, Bush replied, "The idea, I believe, is that you and I would keep George Merck in line, that is, we would see to it that he didn't make too much money or get in jail or anything like that, but on the contrary that he put out a lot of stuff that would cure people's ills. Between your knowledge of George and my knowledge of chemistry I think we could do it and it certainly would be a grand thing for people that had things the matter with them."[50] He warned,

> I'd have to teach you a lot of chemistry, and maybe physiology or something. I don't know how you expect to run George's shop for him unless you get into those things. He makes a lot of queer chemical substances, and people eat them, and that's

> dangerous unless you know what the things are and what they are likely to do to people's innards. Now I understand George himself doesn't know what some of the chemicals are that he makes, just gives them high sounding names, and can't really tell you what their formulas are, and maybe that's all right, for I understand he tries them out on his lawyers before he sells any, and of course he can't get into very serious trouble that way. But somebody in the outfit ought to study a lot of chemistry, and I'm not too sure how good a pupil you are.[51]

Merck could rely on Bush for more than a good laugh. He knew that Bush would support his efforts to enhance the scientific stature of his company and to gain a competitive edge in the marketplace. Bush joined the board of directors in 1949 and served as chairman of the board from 1957 until his retirement in 1962. He played an active role encouraging, not just the top management of Merck, but the entire pharmaceutical industry to support scientific research. At an industrywide conference, he argued, "Its presence [scientific research] in the latter [industrial laboratories] is often highly advantageous, for it tends to lift the tone of the whole effort, and it helps in recruiting if there are nationally known scientists in the organization. Encouraging the freedom of association with scientific societies and of publication . . . renders an industrial laboratory attractive to creative minds."[52]

With Merck at the helm and Bush on the board, Merck & Company made scientific research a fundamental part of their business strategy after the Korean War. Merck was unusual among other firms in the industry, having put their own money into research projects as early as 1933. Even so, research expenses at Merck increased 638.7 percent between 1933 and 1952.[53] At a strategic planning exercise in 1952, Merck resolved to invest even more heavily in research in order to ensure a constant stream of product innovations. Vannevar Bush observed in an internal memo: "The advent of really important new products can prevent the company from being drawn into a situation where it is merely competitive in the manufacture of conventional things, a role for which it is not adapted."[54]

Merck was not the only company pursuing this strategy. By 1953, National Drug was looking for ways to compensate for low returns as well. In a message to stockholders, National's president explained, "it becomes more apparent each year that in order to maintain a successful competitive position in the pharmaceutical and biological markets, the company must rely on a creative and aggressive research department."[55]

A 1955 Arthur D. Little report attests to Merck's extraordinary focus: "Research management now has the full support of the company officials and the directorate, and the importance of research to the company's future progress is probably more generally accepted in Merck than in virtually any other company in the process industries."[56]

Merck's efforts did not go unquestioned. The Arthur D. Little report criticized the company for "overemphasis on work which develops and demonstrates the scientific stature of the organization," implying that these investments were made at the expense of more practical research objectives.[57]

Several board members shared this concern, but Bush urged patience, and explained that in the business of research, one does not pull a rabbit out of the hat every five minutes.[58] He went on to remind them: "Since the war there have been two very large rabbits, namely cortone and B12 . . . every one of them came out of the situation in which we were not only carrying out fundamental research, if you please research of scientific stature, but also because we were thoroughly in contact with others that were doing the same thing in academic circles, and because we were thoroughly alert to the trends in medicine . . . I hope we will not be so short-sighted as to lose our touch with 'development of scientific stature.' "[59]

Merck brought Bush on board not merely to advocate science but to impart war-honed management principles. As one reviewer put it, Bush was "much more than 'wonderful window dressing' . . . He [was] a skilled administrator."[60] In particular, Bush encouraged the OSRD practice of "giving a man his head." Although he believed it necessary to have someone at the helm to provide targets, he maintained that "to tell a fundamental scientist what to work on, or how to go about it, is just about as futile and disastrous as to try to tell your wife what to cook and how to cook it."[61]

Bush argued for a similar balance of power between the CEO and his management committee of vice presidents. He explained, "Every member freely expresses his opinion, but the president decides. Such a relationship can be compared to a general in the field supported by his staff."[62]

Merck also perpetuated the wartime practice of soliciting outside expertise. One former chief operating officer and vice chairman, Antonie Knoppers, maintains that this was one of the most distinguishing features of managerial practice at Merck: "Outside advice played a role in Merck. We were never in-bred, and we always had information from the top people from the outside."[63] This inclination, Knoppers argued, was at the source of Merck's continued success as a leading research and development company.

SOON AFTER BUSH joined the board, Merck & Company decided to enter the vaccine business. In 1953, Merck orchestrated a merger with Sharp & Dohme, a neighboring company with vaccine research and development capabilities.[64]

The business rationale for Merck's move into vaccines was not obvious. Historically, vaccine development had not been profitable. Unlike most vaccine firms at the time, which already had all of their resources tied up in vaccine development, Merck had other lucrative investment opportunities in pharmaceuticals. Part of the motivation may have stemmed from the technological opportunities afforded by improvements to cell culture techniques in the 1950s, which opened the door to large-scale production of a wider variety of virus vaccines.[65] These techniques were being widely adopted throughout industry to produce the polio vaccine, and Merck likely felt pressure to build new competencies in this field. An internal report from this period warned, "If Merck wishes to be a major factor in the field, it seems inevitable that an adequate unit sufficiently versatile to be modified for newer procedures and newer vaccines must be made available before the next major discovery is announced."[66]

While polio vaccines were a widely celebrated medical triumph, cell culture techniques were still new and the future was uncertain. Hilleman contended that Merck had taken a tremendous risk by investing in vaccine research at that time: "It is to the credit of the company that it initiated and funded a biological enterprise at a time when the data base was limited, applicable precedents were lacking, scientists with interest and relevant experience were few in number, and there was little means for obtaining patent protection for discoveries in biology that had been achieved at great cost from pioneering research and development."[67]

Given the risks and uncertainties inherent in the business, it is possible that the decision to invest in vaccines did not rest entirely on the promise of market returns or technological opportunities. According to one historian, there was a looming sense during the postwar period that "breakthroughs in chemical or biological warfare might at any time create an end run around the atomic balance of terror."[68] Merck had already been developing chemical and biological defenses for the military, in the form of nerve gas antidotes and an anthrax vaccine. As Cold War tensions intensified, George Merck and Vannevar Bush could not have been insensitive to the national security importance of vaccines. It is possible that they approached this investment with more than markets on their minds.

The threat of biological warfare weighed heavily on Bush. He wrote, "I am personally somewhat terrified by what may happen on this thing in the postwar world. It is new. But there may be a time when it would be possible to build up by this [biological] means the kind of sudden and devastating attack that would be overwhelming, and the next dictator somewhere who has ideas of conquering the world may see this as a means."[69]

Bush, not one to merely issue warnings, worked with Conant during the war to contain the threat from biological warfare. In a memo to Stimson, they proposed an arms control scheme in which an international body—such as the United Nations—would police biological weapons.[70]

It is likely that Bush did not believe that biodefense was the sole responsibility of the U.S. government. Much like the prewar corporate liberals, he thought that industry had an obligation to protect the public and serve the government. He wrote, "The management of a company and the directors, have a four-way responsibility: to their stockholders, to the public they serve, to their labor force, and to their government. Fortunately there is growing in the country a management group with a distinctly professional outlook and a sense of social responsibility of a high order."[71] Eschewing a straightforward market-driven approach to industrial research and development before a roomful of executives, Bush argued, "There is a genuine need, from a patriotic standpoint, for industry to collaborate in the research effort essential to national defense."[72]

Merck echoed Bush's sentiments on many occasions. He was famous for saying that "Medicine is for the people. Medicine is not for the profits." He regarded his company as "something in the nature of a public trust."[73] Addressing the George Westinghouse Centennial Forum in Pittsburgh, he warned, "Those responsible for our defenses and preparedness in this upset world are alert; they have their programs ready. But they need support—support from scientists, academic and industrial, which should be given generously and in full measure—and it should not wait for an emergency call of patriotism."[74] National security and public health were clearly linked in his mind, and he envisioned an activist role for his company.

THE YEAR 1957 marked a new era for vaccine research at Merck & Company, with the opening of the Virus and Cell Biology Division. Both Merck and Richards had retired two years before, but Bush continued to oversee the company's expanding role in vaccine research. According to Hilleman, who was tapped to lead Merck's new division, Bush "conceived of the ultimate importance of viruses to science and health and requested that Max Tishler, then president of the Merck Research Laboratories, establish a major research program in virology at Merck."[75] Tishler agreed that Merck should make a large investment in biologicals, and, with the support of CEO John Connor and President Henry Gadsen, he began to build the new division.[76]

Tishler recruited Hilleman from WRAIR, where Hilleman had earned a reputation as a talented virologist for his work on influenza and adenoviruses. Hilleman fit Bush's ideal of a "science integrator" exactly. He was brilliant, multidisciplinary, and self-directed. Hilleman thrived at WRAIR, where he was given free rein to direct his own research. However, it was not clear that he would fit neatly into Merck's corporate culture.

Reflecting on his early introduction, Hilleman explained, "At Merck I was like a big wart on an elephant's butt."[77] He refused to attend meetings that did not pertain to his research. Once when he was confronted by a superior about his unorthodox approach to administrative duties, Hilleman informed him, with characteristic tact, "If I wanted an administrative job, I would be your boss by now."[78]

Fortunately for Hilleman, Bush was outspoken about the importance of granting freedom within the bureaucracy in order to retain talent. Bush often argued, "The primary need is to find the really great research scientists with an intense interest in parts of the problem and see to it that they have every possible support as they work, on their own ideas and in their own manner, without administrative interference or control of any sort."[79] The company followed this dictim and granted Hilleman managerial control and budget authority to pursue vaccine projects according to his own vision.

Hilleman's vision was shaped by his experience in military labs and he began to introduce some of the same integrated research practices that had served him so well at WRAIR. While working on the adenovirus vaccine, Hilleman was "involved in every step in the process," including the earliest identification and isolation of a new virus.[80] He recalls, "I went on epidemiological trips at the drop of a hat. I always had a bag packed and was ready to go by plane or train or whatever it required. If there was a respiratory disease that looked interesting, I would go out there and take a look, collect the specimen, see the patient."[81]

Hilleman first isolated adenoviruses from military personnel in 1954.[82] Several years later, Hilleman and his associates on the Commission on Acute Respiratory Diseases of the AFEB traced a causal relationship between the presence of adenoviruses and acute respiratory disease among the military recruits.[83] Hilleman found that going back and forth between the field and the lab facilitated vaccine development because it allowed him to forge links in the chain of specialized epidemiological, clinical, and laboratory-based knowledge necessary to develop a vaccine. In this role, he offered a one-man solution to many of the technology and knowledge transfer problems that can thwart vaccine development initiatives.

At Merck, Hilleman created organizational structures and routines to support the integrated research and development methods he honed at WRAIR. According to former Merck CEO Roy Vagelos, "All you had to do was walk into his [laboratory] to realize that he had transformed [the department of] virus and cell biology at Merck into a military organization. Everybody knew what to do at the beginning of every day."[84] He also created "crossover" teams to integrate the expertise of members from different departments. These teams were designed to "permit knowledge to be passed as necessary from the immunology to the chemistry to the physiology department, etc."[85]

Hilleman explained that, unlike most other companies at the time, "we did field investigations, epidemiology, from clinic to clinic, isolated viruses to attenuate. We had the complete spectrum."[86] This capacity for interdisciplinary problem solving set Merck apart from their friends in academia. Tishler noted, "Sometimes university researchers [came] to us for help on something that could be solved by a fellow on the floor below their own lab."[87]

Hilleman explained why he chose to organize research in this manner:

> I've found it to be of extreme importance and help to have a concentrated program of very great breadth, covering a diversity of diseases and extending from the laboratory bench through the clinic. By this I mean a program ranging from discovery of the cause of disease and the assessment of its importance through the development of means for control and the proof of safety and efficacy of the method in large-scale clinical studies in man. This provides for self-catalyzing spillover from one study to another and encourages progress toward the target objective without the disruptions imposed by

> passing the successive stages of development from one research group or department to another.[88]

As was the case with World War II vaccine development programs, this interdisciplinary approach required a top-down management structure to integrate work streams from a wide range of participants. At Merck, as at WRAIR, Hilleman "was the main coordinator and link in the chain . . . I did all the field trials, I was involved in every step and every decision . . . It was directed coordination."[89] This practice permitted quick decision making and enabled efficient operation. It was also, in his experience, a necessary evil: "Researchers like to deviate," he commented, "chasing every which thing that comes their way. I could walk out of the lab for one week—I'd come back and I'd already find these bastards working on something else. When you're gone three weeks, you hardly know the laboratory anymore. You need to crack the whip."[90]

As with all dictatorships, the quality of governance is a function of the energy and intelligence of the individual in charge. Luckily for Merck, Hilleman was a brilliant force of nature. While he was the director of Merck's Virus Cell and Biology Division from 1958 to 1984, Merck introduced forty-one licensed vaccine products to the market. Six (the meningitis, measles, mumps, rubella, hepatitis B, and varicella vaccines) were entirely new.[91] "That system worked," he concluded.[92] "Everybody was happy. They were overjoyed to be productive. They could see these vaccines coming out."[93]

WHILE THE WORLD WAR II legacy of personal friendships, ideologies, and research practices at Merck was somewhat unique, WRAIR-based collaborative networks extended to other industrial labs as well. The community of skilled research scientists was small, so it was not unusual for individuals to know one another from a prior stint at WRAIR. Common training and mutual respect between current WRAIR researchers and WRAIR alumni in industry facilitated collaboration between the two groups. They often shared research results, biological samples, and laboratory equipment. An enduring sense of duty in the postwar period further encouraged collaboration and industry often felt compelled to accept vaccine development projects for the military. These combined characteristics of familiarity and indebtedness are unique to this period and evident in the postwar developmental history of the meningitis and influenza vaccines.

The military scored their first success against meningococcal meningitis during World War II. According to James Simmons, the prophylactic use of sulfadiazine within the armed forces dramatically reduced the incidence of meningitis. In 1944, he reported, "The case fatality rate [from meningitis] which was 38 percent in the previous war has been less than 5 percent."[94] By the early 1960s, however, military doctors began to see patients infected with sulfadiazine-resistant strains of *Neisseria meningitidis*. Recruitment camps, in particular, began to suffer from outbreaks of bacterial meningitis that did not respond to antibiotics. Occasionally, with no other recourse, the military was forced to close down camps to stem outbreaks. After one particularly severe outbreak among recruits in 1963, the AFEB initiated studies within military populations to determine which strains of the bacterium predominated.[95] Within a year, the army had set up a new research unit at WRAIR to study the problems of vaccine development.

Malcolm Artenstein, a career military physician who had isolated the rubella virus with Paul Parkman and Edward Buescher in 1962, headed the new unit. During this time, the army was able to call up draft-eligible research scientists. Taking advantage of the draft, Artenstein hand-picked a recent graduate from NYU School of Medicine, Emil Gotschlich, to help develop a vaccine. Together, Artenstein and Gotschlich decided to revisit Elvin Kabat's research, which had been conducted at Columbia University in the 1940s, before the widespread availability of antibiotics had made the search for bacterial vaccines seem unnecessary.

Kabat had identified immunogenic properties of bacterial polysaccharide capsules, which suggested that it might be possible to induce active immunity against meningococcal meningitis with a polysaccharide vaccine. Building on this research, Artenstein and Gotschlich—along with Irving Goldschneider, another recent draftee—began to isolate and identify polysaccharides in the capsules of *N. meningitidis*. Their research demonstrated that these polysaccharides could generate antibodies to meningococcal types A and C, the strains most commonly found in U.S. recruitment camps. By 1969, they developed pilot lots of this vaccine and began to test it for safety and efficacy on army recruits.[96]

Once the WRAIR team demonstrated the feasibility of a meningitis vaccine, the army began to solicit offers from industry to manufacture the vaccine on a larger scale. The contract had first gone to Squibb, but they could not figure out how to obtain high yields of polysaccharides. After working unsuccessfully for two years to scale up production of the vaccine, Squibb pulled out and WRAIR contacted Tishler at Merck.

When Tishler approached Hilleman with the military's request, he sent the scientist into a tailspin. "At the time that I came to Merck," Hilleman explained, "I didn't want anything to do with bacteria. Army officers talked to Max Tishler, trying to get him to take the contract. Max then came to me and I said, 'Meningicoccus? Christ! That's a bacterium, isn't it? I don't know anything about bacteria. I've just heard of them.'"[97]

Having just developed the measles, mumps, and rubella vaccines, Hilleman was anxious to continue work on new viral diseases. He knew little about bacterial diseases and the sporadic nature of the disease did not bode well for efforts to test or sell a meningitis vaccine. Yet his sense of loyalty and indebtedness to the military ran deep, and he felt obliged to comply with their request. "I thought, oh my God, this would be a real disloyalty to the military so I said, oh OK, I'll take it over."[98]

Before Hilleman accepted the contract, two scientists at the National Drug Company in Swiftwater, Pennsylvania, had already started to investigate this project. This work was directed by James Sorrentino, who, like Hilleman, had trained at WRAIR and had recently transitioned to a job in industry. From 1960 to 1967, Sorrentino had been a research scientist in WRAIR's biologic group under Joseph Lowenthal.[99] Sorrentino's training and experiences were not unlike those of Hilleman, who had been director of WRAIR's virus and respiratory disease division in the 1950s. Sorrentino explained, "I learned every aspect of vaccine development. I never could have done that anywhere else because I would have been pigeonholed as a development guy, or a research guy or a manufacturing guy. I learned the animal part of evaluation of vaccines in terms of efficacy and safety. I learned the manufacturing aspects of vaccine development, the research aspects of what comes before and after vaccines are developed . . . it was unique training."[100]

Sorrentino, like Hilleman, was frustrated to discover that, in industry, "everyone was a specialist." He explained: "They didn't know what went before or after their piece of it. You get bias by the researcher and the manufacturer. They have their own agendas and it gets in the way of smooth development of the drug."[101]

In an effort to overcome these departmental separations, Sorrentino forged an alliance with Don Metzgar, who had joined National a year earlier, after spending five years working with Hilleman at Merck.[102] Metzgar was hired as a senior virologist and Sorrentino had been hired to manage vaccine development and manufacturing operations. Within months of Sorrentino's arrival, the two were sharing offices and, to a certain extent, a budget, as they began to coordinate work across research, development, and manufacturing. According to Sorrentino, "We had those three aspects and so the only thing we had left to control was quality control; so we did it by brute force."[103]

The first project they chose was to develop a highly purified influenza vaccine. Prior to 1970, influenza vaccines were not widely used, in part because they were highly reactogenic. Sorrentino recalled, "Heretofore, you had a flu vaccine that took your arm off."[104] In the 1960s, influenza vaccines were purified, using a Sharples centrifuge, with methods developed under CMR contracts in World War II. Summarizing what was state of the art for influenza vaccine production at the time, Sorrentino explained, "All they did was spin down the egg material, took the sludge, re-hydrated it, diluted it, put a preservative in it, and shot it into your arm."[105]

While at WRAIR, Sorrentino worked on an eastern equine encephalitis vaccine. His search for ways to purify the virus introduced him to new density gradient techniques developed at WRAIR. It occurred to Sorrentino that he might use these techniques to further purify influenza vaccines if he could find a way to get the virus to band according to its density gradient. By purifying the influenza vaccine in this manner, he could reduce reactogenicity and improve demand for the vaccine.

Charlie Riemer, at Eli Lilly, had recently published a paper describing how his team had adapted a high-speed gas centrifuge, originally developed by the Atomic Energy Commission to enrich uranium to purify viral materials.[106] Inspired by this paper, Sorrentino set out to devise his own method for separating viral material from the allantoic fluid in a more purified form.

National did not have an ultracentrifuge and they were unwilling to invest in one to test Sorrentino's idea. Undeterred, Sorrentino tapped his network of military research scientists for assistance. He went to see Norman Anderson, director of the Molecular Anatomy Program (MAN) at Oak Ridge National Laboratory, who had also adapted an ultracentrifuge to work with biological agents. Sorrentino was thrilled when Norman Anderson just "opened his lab to me." He recalls, "I would go out and seek help from the government and the program at Oak Ridge and get ultimate collaboration."[107]

He began to fly down to Tennessee regularly, with his influenza samples on the seat next to him. He ran them through the government's centrifuge and peered at the results through the government's electron microscope—unencumbered by licensing fees, material transfer agreements, or any other intellectual property concerns.

In short order, Sorrentino was able to demonstrate that the process worked. With this data in hand, he convinced National to invest in the new technology and facilities required to manufacture the vaccine. Once National put this new vaccine on the market, Sorrentino recalled, "We ended up forcing the entire industry to purify their product."[108]

Soon after developing his purified influenza vaccine, Sorrentino was restless for a new project and anxious to try his purification techniques on a new vaccine. He recalled from his time at WRAIR that "Meningitis was one of those things on the burner that needed to be done."[109] Sorrentino had just joined the biologics research group at WRAIR and started work on a meningococcal vaccine before leaving for National. Sorrentino called Sanford Berman, another former colleague at WRAIR, to check the status of the project.

Berman relayed that WRAIR had taken the development as far as they could but that the vaccine required further purification. Thus far, no one had managed to produce it commercially. Sorrentino agreed to try immediately. This time he had little trouble getting National to sign off on his project. "After I gave them flu," he boasted, "I could come to them with anything. I was like Maurice Hilleman to National Drug."[110]

During the 1950s, 1960s, and early 1970s, vaccine development was unfettered by intellectual property or liability concerns and military-industrial collaboration was characterized by a freewheeling, informal collegiality and trust. When National agreed to develop the meningitis vaccine, Sorrentino claims that no contracts, licenses, or patents changed hands: "It was free-exchange. I didn't sign a single paper."[111] Metzgar recalled the simplicity of the arrangement with amusement and chagrin: "Jim brought back this bottle of paste [WRAIR's seed stock of *Nisseria meningitidis*]. We had no authorization and no budget."[112]

Familiarity with personnel in WRAIR's biological research division significantly facilitated National's efforts to develop the military's vaccine. Sorrentino and Metzgar soon discovered, however, that technology transfer, even under the most favorable conditions, is rarely straightforward. Artenstein and Gotschlich had published their work on the vaccine, but "Like everything in the literature, when you start trying to follow what they say in the literature, it is not exactly reproducible."[113] Whenever Sorrentino ran into a roadblock, he merely had to "pick up the phone for Joe Lowenthal and say, 'Hey Joe, I've had a problem with this step.' "[114] Similarly, Sorrentino and Metzgar had full access to Gotschlich, who on several occasions gave them firsthand advice on how to translate WRAIR's research in their own labs at National. Even so, they found that it took months merely to adapt what Artenstein's group had done at WRAIR.

Once they were finally able to reproduce WRAIR's results, Sorrentino and Metzgar were confronted with the problem of scaling up pilot lots. Sorrentino explained, "It's good for the manufacturer to know what the developer does, but there are a lot of things you have to change when you start scaling up. A lot of things that aren't as efficient and that aren't as sensitive; yields are different. So there have to be a lot of changes that are made in the process."[115]

One such change consisted of learning how to grow *Nisseria meningitidis* in large fermenters. *Nisseria meningitidis,* Metzgar recalled, was a fussy organism, and growth rates were highly sensitive to media construction, temperature, and other environmental conditions.[116] To some extent, Metzgar explained, it was never reduced to an exact formula: "As a matter of fact, we are still working on the yields of that product."[117]

Nonetheless, Sorrentino and Metzgar made tremendous headway. They devised methods for separating the polysaccharide capsules from the bacterial cells and distinguishing among them by size. Applying Sorrentino's density gradient techniques to larger polysaccharide molecules proved difficult at first. He recalled that this particular problem stumped them for about a year until they devised a way to exclude molecular sizes on columns according to molecular weight using gel permeation chromatography. With this method, Sorrentino and Metzgar were able to select the larger molecules, which elicited a better antibody response for their vaccine. Once they cleared this hurdle, they were poised to manufacture a large quantity of vaccine for clinical trials.

Proving efficacy presented another problem. Sorrentino recalled that meningitis attacked two in every fifteen thousand people in the United States at that time. They would have to immunize hundreds of thousands of subjects to do an efficacy trial of any statistical significance. To circumvent a large, expensive and time-consuming trial, National entered into an agreement with the South African health minister to conduct clinical trials in gold mines, where the disease was both endemic and epidemic. "If you go down in the gold mines 2,000 feet," Sorrentino explained, "it is an ideal environment for transmitting. We went from a very low rate to 15/1,000 people that come down with meningitis in the mines."[118]

Meanwhile, the army was anxious to provide efficacy data to the NIH's Bureau of Biologics (BOB) to get the vaccine licensed as soon as possible. The BOB arranged to pool efficacy data from Merck's and National's clinical trials to halve the time required for clinical testing.[119] Sorrentino explained, "It was something that happened back then because we were very collaborative."[120] Metzgar further attested to the collegial atmosphere surrounding the project, recalling, "Jim had come from Walter Reed and knew the people there, and I came from Merck. I knew who the players were—who I could pick up the phone and call if I had a problem with something."[121]

The meningitis vaccine was a tremendous technological and clinical success. It offered safe and effective protection against a deadly and previously unpreventable disease. Subsequent tests revealed that Merck's vaccine was effective in the most vulnerable age category: very young children.[122] There was also strong global demand for the vaccine, but high demand did not translate into high profits for industry. The greatest demand came from countries such as Brazil and Africa, who could not afford the vaccine at full price. The majority of U.S. sales went to a travelers' market, which was quite small, and to low-margin military contracts. The rest was donated to countries suffering a high incidence of the disease.

Despite the technical success, clinical effectiveness, and medical need for the vaccine, Merck eventually decided that meningitis outbreaks were too few and far between in their primary market to justify further investments in producing the vaccine.[123] National continued to manufacture it and suffered low returns for decades until U.S. civilian markets slowly caught up with military needs. Demand for the meningitis vaccine has increased in recent years, as antibacterial resistance has spread more widely throughout the general population and as meningitis outbreaks on college campuses have become common.[124] The military was, once again, a premature but accurate indicator of an emerging public health threat.

Neither Merck nor National (then a division of Richardson-Merrell) made a strict business decision when they decided to develop a meningitis vaccine in the early 1970s. The project reflected a duty-driven spirit of vaccine development that existed within some firms during the postwar era. During this period, it was not unusual for some executives to consider their public duties, alongside their fiduciary duties, as George Merck and Vannevar Bush often did. These firms had more freedom to make vaccine development decisions that reflected a wider range of values and long-term development plans than they did later when companies began to focus more intently on annual, and even quarterly, earnings. The WRAIR alumni who staffed Merck's and National's scientific divisions reinforced this sense of obligation to the military, and they supported military requests to develop vaccines even when there was not a compelling business rationale.

Industry's willingness to develop a meningitis vaccine for the military suggests that, on occasion, Merck's and National's vaccine divisions operated according to the principles of what anthropologists and sociologists call a "gift regime."[125] Economic anthropologists distinguish between gift and commodity exchanges as a way to track the shifting social and moral identity of objects exchanged within a variety of cultural contexts.[126] Whereas commodity exchanges exact a price, gift exchanges are personal and exact a sense of social debt.[127] The former operate in a market of arm's-length transactions, whereas the latter operate in the context of shared experiences and intermingling personal histories of the sort observed among veterans of wartime development programs and WRAIR alumni.

The personal connections that sustained this culture were particularly strong at Merck & Company, where a number of OSRD members, including Vannevar Bush and Alfred Newton Richards, followed George Merck into the pharmaceutical industry. Personal connections to military research labs were reinforced as WRAIR developed a reputation as a center of excellence for infectious disease research and as companies began to seek collaborative partners and new hires at WRAIR.

Bruce Smith observed that the political and cultural landscape of the postwar era was characterized by "patriotism and broad public support, deference to executive leadership, and the subordination of partial interest to the larger national interest."[128] Military planners, industrialists, and research scientists recognized the value of vaccines to national security and public health. Industrial leaders who had participated in World War II research and development programs were predisposed to invest in research to accept military contracts for the public good. Early Cold War anxieties reinforced this predisposition, driving military and industrial investments in vaccine research. For a time, vaccines enjoyed the economic and social status of a public good, or of a "gift" granted out of a sense of social obligation. As such, vaccines were not always subject to a straightforward, rational calculation of costs and returns in the postwar period.

According to one internal study, Merck concluded that "the reasons [why companies stayed in the vaccine business were] mostly NOT of the 'rational strictly business' type. The major one for many [was] political."[129] Merck was no exception; Vannevar Bush and George Merck accepted military contracts and urged the rest of the industry to uphold their public duty and do the same. Similarly, Hilleman accepted the army's meningitis contract not because he thought this vaccine would be profitable, and not because he had a professional interest in bacterial polysaccharide vaccines, but because a failure to do so would be disloyal. Sorrentino described National's attitude in a similar vein: "It may sound syrupy, but they [National] really believed that they had something special and if they could give it to the government, they would. No matter how little money they made, they felt that if the government needed it, they had to respond to that. They were always working around the clock. I can't tell you how many times I saw those old Army trucks pulling to the back of those bays to pick up a supply."[130]

Duty-driven postwar collaborative networks between WRAIR and industry catalyzed innovation in part because they maintained an important link between users and developers. These informal networks replaced the formal arrangements that had facilitated the translation of research into products during World War II. The migration of talent from WRAIR to industry sustained an important community of practice for vaccine development and ensured that the research practices of "science integrators," such as Maurice Hilleman and James Sorrentino, were adopted by industry. As development of the meningitis and influenza vaccines revealed, integrated tactics, combined with a sense of personal obligation and loyalty to the military, had a significant and positive influence on industrial vaccine innovation throughout the postwar period.

5
The End of an Era

By 1979, a highly productive era of vaccine innovation was drawing to a close. Many manufacturers had either exited the vaccine business or seriously considered getting out. While several factors discouraged private sector vaccine investments at this time, industry observers often credit the swine flu affair with setting a wave of industry consolidation into motion.[1]

In 1976, an army recruit at Fort Dix died of an upper respiratory ailment after an overnight hike. The New Jersey Health Department isolated an unusual strain of influenza that was normally found in pig populations from this soldier and many others at Fort Dix. Ever since the 1918 pandemic, virologists had feared a reoccurrence. Since both antigenic proteins on the surface of this virus were significantly different from any known viral antigens circulating at that time, some feared that a pandemic was imminent.[2] After a series of urgent meetings and heated debates, the federal government proceeded with a program to develop and administer a swine flu vaccine to the entire U.S. population before flu season returned in the fall and winter of 1976–1977.[3]

Vaccine producers agreed to assist the government with this campaign and launched programs to produce record quantities of vaccine on short notice. Some manufacturers reportedly "committed their manpower, plants and working capital to the operation before pinning down profit and loss factors."[4] These companies (Parke-Davis, Merrell-National, Wyeth, and Merck) did, however, demand liability protection from vaccine-related injury claims. As with all vaccines, the swine flu vaccine was likely to produce adverse reactions in a small but certain percentage of the population. Because this vaccine was going to be administered to everyone, the overall number of adverse reactions was bound to be higher.

In June of 1976, the director of the American Insurance Association announced that they would not provide coverage for swine flu producers and that any preexisting coverage would be terminated on June 30.[5] Manufacturers held that without liability protection they could not shoulder the risk of participation.[6] Faced with the impending collapse of the immunization program, President Ford urged Congress to draft legislation that would allow the government to provide liability protection so the program could proceed. Congress passed legislation that offered companies liability protection on the condition that they did not profit from swine flu vaccines sold to the government. Companies upheld their agreement to produce the vaccine, despite the no-profit clause.[7] Immunizations began in October.[8]

Within months of the first vaccinations, there was no evidence of a pandemic, but a small number of vaccinees developed Guillain-Barre syndrome, a disease affecting the peripheral nervous system. Concerns about this rare side effect prompted the CDC to terminate the program. While industry did not suffer directly from product liability suits filed against the swine flu vaccine, publicity about the hypothesized adverse effects of the vaccine raised public awareness of the risks associated with vaccination in general.

Ironically, public attitudes toward vaccination were taking a negative turn just as a combination of vaccination and sanitation campaigns had brought the incidence of infectious disease to an all-time low in the United States.[9] With few salient reminders of disease, individuals began to worry more about the potential side effects from a vaccine than the disease it prevented. Don Metzgar marveled at this generational shift, which strongly contrasted with his own experience growing up: "There were no vaccines. There was polio running rampant and people can't relate to that any better than they can relate to World War II."[10]

Public concern about vaccine safety launched an avalanche of product liability suits for other vaccines not protected by the indemnification act.[11] Merck recorded annual increases in the number of vaccine-related lawsuits from fifteen filed in 1979 to a total of forty-seven in 1986.[12] Merrell-National also felt the sting of what they called "a national crisis of product liability."[13] In 1985, Connaught (which had acquired Merrell-National) reported that a total of $5 billion in claims had been filed, with $3 billion in lawsuits filed against manufacturers of the combined diphtheria, tetanus, and pertussis vaccine (DTP).[14]

The growing risk of product liability suits strained an industry that was already struggling with low profits. Merrell-National complained that senior managers "were spending more than half of their time on a business segment that only amounted to about 5 percent of their revenues."[15] Frustrated by low returns, Merrell-National decided, along with half of the industry, to divest its vaccine division.[16] Of the companies that remained in the business, many, such as Merck, stopped manufacturing DTP and influenza vaccines altogether and scaled back on other vaccine research and development investments.

It is often argued that product liability suits strained budgets and accounted for declining rates of innovation and supply in the 1970s. Product liability was not a new phenomenon, however. Courts began awarding large sums to individuals claiming vaccine injuries by the early 1960s, when vaccine innovation was still relatively robust.[17] Low profits were not new either. A 1979 report stated that industry had long suffered from "big capital investment requirements, complicated manufacturing processes, burdensome licensing procedures, a static market for established vaccines, and, until recently, an overabundance of competition."[18] This same report noted that "in more than 10 years, Merrell National, one of the nation's largest pharmaceutical companies, is reported to have failed to turn a profit in vaccines."[19] And yet, these conditions did not prevent this company from investing in ambitious meningitis and flu vaccine projects in the 1970s.

While the economic argument for vaccine investments changed somewhat in the wake of the swine flu affair, the political argument changed dramatically. The swine flu experience damaged the reputation of immunization campaigns and overturned a belief that the social obligation to respond to government vaccine requests (and the associated public relations benefits) outweighed the financial costs. This belief—a descendent of the corporate liberal movement in the 1930s—was forged during World War II and honed in the postwar years, but it did not survive the swine flu affair.

When the federal government announced that it would vaccinate the nation against swine flu, Robert Hendrickson, director of Merck's manufacturing at the time, recalled, "Suddenly we got hit with this tremendous requirement for millions of doses—in a very short time span. So we geared up like mad to try to do this thing."[20] Hendrickson explained that developing and scaling-up the vaccine for high-volume production was an extraordinary accomplishment undertaken for no profit. Yet "it ended up engendering nothing but bad will despite a really massive effort to try to do something for the benefit of the country . . . We got nothing but criticism from the whole program because, as it ended up, the vaccine wasn't needed . . . You start out with something which is being done as a goodwill gesture for the benefit of mankind, and like many things that happen in the industry, it gets turned against you because of suspicion that you did something wrong, that you put out a harmful product."[21]

As public relations benefits evaporated, many companies cast off their vaccine divisions to focus on more profitable product lines. Merrell-National, Eli Lilly, Pfizer, Parke-Davis, and Dow Chemical were among the large U.S. firms to exit the business. By 1985, only Merck, Wyeth, Lederle, and Connaught remained.[22]

After years of industry consolidation, economic prospects for the remaining players improved considerably. Merck held 60 percent of the U.S. market for vaccines, Lederle 18 percent, Connaught 14 percent and Wyeth 8 percent.[23] Each company also held quasi-monopolies over individual product lines, which gave them the freedom to raise vaccine prices.

Reflecting on business in the 1980s, Metzgar recalled, "Ironically, Connaught began to make money when litigation became so prevalent."[24] He explained: "[The company] started to charge money to cover self-insurance needs and raised prices. There was a period of time when we didn't sell anything—almost drove ourselves out of the market. Eventually, other companies began to look at what we were doing and followed suit."[25]

In an effort to stem price hikes and to prevent more manufacturers from exiting the industry, Congress passed the 1986 Vaccine Compensation Act to relieve manufacturers of liability for nonnegligent vaccine-related injuries sustained in compulsory immunization programs.[26] This legislation reduced product liability costs (compared to the 255 DTP suits filed in 1986, only four were filed in 1997), but vaccine prices continued to rise.[27] From 1980 to 1995, vaccine prices rose at a greater rate than the Consumer Price Index or the Producer Price Index.[28] Lance Gordon, vaccine research scientist and CEO of the ImmunoBiologics Corporation, explained, "When I entered commercial vaccine development in 1980, the aggregate U.S. market for the core pediatric vaccine, DTP, was approximately $6 million, a market that was shared by nine manufacturers . . . [After 1986], sales of the DTP vaccine in the U.S. had grown to almost $300 million."[29] Gordon claimed, "The larger sales were due almost entirely to product liability price increases." One study attributed price hikes, in part to the rising cost of research and product liability, but it concluded that a large portion of the price hikes were consistent with monopoly pricing behavior and reflected the growing market power of remaining vaccine producers.[30] Initially, price hikes may have reflected self-insurance needs, but after 1986, they began to generate profits.

Before the market improved, Merck had also considered exiting the business in 1979. Roy Vagelos, director of Merck Research Laboratories at the time, recalled a gloomy internal planning report: "I couldn't argue with the figures, which were correct. One curve showed the total revenue from vaccine sales and another the costs of

Hilleman's vigorous research program. The two lines were drawing together ominously for the future of Merck vaccines."[31] Ultimately, Merck decided to stay in the business because, according to Vagelos, it "had a well-deserved reputation for social responsibility. It was the major America vaccine innovator and producer, and it had a powerful obligation to carry forward in that role and to try to make the business profitable."[32]

In reality, however, Merck began to treat vaccines like any other business opportunity. Douglas MacMaster, president of Merck Sharp & Dhome in the mid-1980s, warned Congress that even though Merck decided to stay in the business, the days of Merck's industrial patriotism were over: "The vaccine business is a high technology business, not a commodity business, and profitability is not determined by a sales minus costs basis. Profitability is determined by assessing our total business situation, looking at the elements affecting our total business including cost of promotion, capital expenditures, production costs, costs of raw material and the cost of research and development."[33] In a carefully worded statement, he explained, "Merck has a long standing commitment to vaccine research. We believe it is a matter of public trust to remain in the marketplace as long as possible. Nevertheless, the extent to which we are able to allocate resources to the development of new vaccines is certain to turn in part on the profitability of such products."[34]

The manner in which Connaught (the new owner of National Drug's manufacturing facility) chose its next vaccine development project also revealed a new emphasis on market-driven vaccine development. In contrast to the duty-driven spirit in which National assumed work on the influenza and meningitis vaccines in the early 1970s, Connaught decided to develop a *Haemophilus influenzae B* (HIB) vaccine strictly on the basis of market potential. Marketing director Doug Reynolds commissioned Rutgers University to conduct a survey asking physicians whether they would routinely administer meningococcal meningitis vaccines in their office practice. The survey indicated they would not, but that they would be interested in a vaccine to prevent HIB infections (another source of meningitis) in young children.[35] As a result, Connaught promptly dropped all military contracts to develop a vaccine against group B meningococcal meningitis and started working with the NIH to develop HIB vaccines.

Once industry executives began to evaluate vaccine development projects on their commercial merits alone, industry scientists had less freedom to pursue their own projects and serendipitous opportunities for military-industrial collaboration diminished. Metzgar recalled that "prior to that time, vaccines were a sideline. They were not market-driven research investment decisions."[36] He explained: "Corporate willingness to collaborate relaxes when the product potential for a reasonable profit does not exist. This mentality of small revenue and community service prevailed when vaccines were a boutique enterprise, whereas now vaccines are a mainstream revenue producer for manufacturers that are still in business."[37] Metzgar observed that the "most productive collaboration happened when the vaccine business was less than 5 percent of the revenues of National Drug."[38]

~ POLITICAL AND SOCIAL transformations in the 1970s encouraged the public sector to underinvest in vaccines as well. During the Cold War, the DOD supported international vaccination campaigns to win hearts and minds and built international epidemiological networks to detect new and emerging diseases that posed a threat to military personnel. As the United States began to pull out of Vietnam and other contested countries, the military's commitment to these strategies began to slacken. The DOD began to cut back on its international laboratory network in particular, and "epidemiological surveillance in these countries slowed to a crawl."[39]

These labs had furnished the military with uniquely strong capabilities in epidemiological field operations and clinical trials because they allowed basic research scientists and field investigators to work side by side where diseases of military significance were endemic. The Walter Reed Army Institute of Research (WRAIR) has long recognized that "one must go to the source of the problems in order to study them."[40] According to a memo to the director of WRAIR, "supporting laboratory services [were] best brought in as close as possible to the actual site of work rather than left to be maintained at a long distance where transportation problems, perishability of shipments over long distances, inadequacy of communication, etc., all play their part in reducing the efficiency of the field operation . . . To take one specific example," the memo noted, "studies on cholera infection in mice have been carried out for years in all parts of the world, and have contributed practically nothing to our understanding of the nature of the disease . . . combined American and Thai teams working in Bangkok last spring resulted in some major advances in our knowledge of certain aspects of cholera, and this is only a small beginning achieved in a two months' period."[41]

Despite the advantages of this system, in 1979 Congress considered handing the DOD's international laboratories over to civilian contractors, if not eliminating them entirely. While the DOD managed to retain some international lab capacity, Philip Russell, former chief of virology at the U.S. Army SEATO lab in Thailand, observed that budget cuts slowly starved the remaining labs of critical resources: "There is some strength left in those labs, but they don't have the same freedom of action and capability that they once did."[42]

The fate of offensive biological programs during the 1970s was even more dramatic. President Richard M. Nixon unilaterally renounced the development, possession, and use of biological weapons in 1969, and the United States began to dismantle its offensive programs as the Biological Weapons Convention opened for signature in 1972.[43]

Nixon did not eliminate biological warfare programs because they were ineffective. To the contrary, a series of open-air tests conducted in the Pacific near the Johnston Atoll and Eniwetok Atoll demonstrated that pathogens could be effective mass casualty weapons.[44] Biological weapons were poor deterrents, however, and they had limited retaliatory value due to their delayed effects. Furthermore, after a decade of biological weapons research, investigators rediscovered what World War II biodefense scientists already knew: that offensive measures were far easier to devise than defensive measures. Whereas offensive weapons could be formulated in two to three years, effective countermeasures (usually vaccines) took anywhere from seven to ten years to develop. Some scientists, most notably Harvard biologist Mathew Meselson, were concerned that the U.S. offensive program was generating more problems than it solved.

Nixon's decision did not prohibit the DOD from investing in defensive research, but it did limit progress in the field. In 1971, General William Creasy, head of the U.S. Army Chemical Corps, argued that it was difficult to pursue a robust defensive research program without offensive research. Testifying before Congress, he explained, "You cannot know how to defend against something unless you can visualize various methods which can be used against you, so you can be living in a fool's paradise if you do not have a vigorous munitions and dissemination-type program."[45] Given the interdependence of offensive and defensive research, the abolition of the former was felt by the latter. The Pentagon slashed the biological research and development program budget in half and many military-sponsored vaccine development projects lost momentum.[46]

Other political and social shifts began to limit federal support for DOD-sponsored biomedical research during this period as well. During the Vietnam War, politicians began to question Vannevar Bush–inspired notions that federal investments in research and development served the public interest. It was no longer as accepted, as it may have been in 1951, that "government agencies represent the will of the people."[47] According to Daniel Kevles, "science" became guilty by association with an unpopular war. As the war escalated, he argued, "critics turned to searing attacks on science for its close identification with the military, including its advisory relationships to the armed services and DOD's salient presence in academia and training . . . by the late 1960s, the dissidents had produced a coalition that brought a halt to the geometrical growth rate in federal science (while the federal budget had risen elevenfold since 1940, the research and development budget had exploded some two-hundredfold)."[48] These attempts were fairly successful, he concludes, since "by the mid 1970s, the federal R&D budget, then about $19 billion, had fallen in constant dollars about 20% below what it had been in 1967."[49]

Political pressure to curb federal research and development expenditures led Congress to amend the military procurement and research authorization bill of November 1969 with what became known as the Mansfield Amendment. This legislation prohibited the DOD from financing research not directly related to a military function or operation. Although the language was modified a year later to permit appropriations for a wider range of research activities, this amendment reflected a growing predisposition to roll back federal support for non-weapons-related research in the DOD.

In 1969, Congress began to cut through a wide swath of the military medical research budget.[50] As the United States began to lose the Vietnam War militarily, the Armed Forces Epidemiology Board (AFEB) discovered that military medical research was beginning to lose politically. One member noted, "A few years ago, Vietnam was a good word. Now with the de-escalation over there, what is the objective?"[51]

In 1971, the DOD announced that it would conduct a management survey of AFEB operations. As rising oil prices, international industrial competition, and stagflation trampled the U.S. economy, the AFEB was only one of many government organizations undergoing budgetary review in the early 1970s. However, there was also a new movement within Congress to eliminate conflicts of interest within the government. The survey concluded: "The present system violates the spirit, indeed if not the letter of the law. It is improper to hold a government contract and be an official member of the review group that technically approves one's research proposal even if the advisor leaves the conference room during the discussion."[52] On the recommendation of this report, the AFEB was stripped of its research function and reduced to an advisory role.

Firsthand familiarity with sponsored research was, in the words of one WRAIR alumnus, "a reason that the AFEB system was so successful in the early days." He conceded, however, that by 1972, "the conflicts of interest inherent in the system were too obvious to ignore."[53] Other members felt this ethical gain came at the expense of progress in the field as well. At the annual AFEB meeting in the following year, one speaker noted: "With the demise of the Commission as we've had it in the past, we are really faced with a paradox. The field state of the art is getting to the point where I think rapid and important advances could be made on a modern biological basis. Never before have we had the tools that we have right now. At the same time, we are losing the one major organization in this country that has held the field together and has provided the major stimulus for all of this."[54]

Budget cuts within the AFEB, and the Military Medical R&D Command more generally, dovetailed with shifting research priorities. Colonel Greenberg from the Preventive Medicine Division of the Army reported: "At the present, all of the infectious diseases are in competition with environmental pollution. Federal legislation and congressional interest now requires us to give a lot of attention to this."[55]

Vaccine research began to suffer at WRAIR as well, but for reasons that had less to do with budget cuts than with a growing collection of institutional and professional factors. Chief among these was the termination of the doctor draft after the Vietnam War. An internal army report from 1974 warned, "Filling the junior staff positions, although adequate at the moment, is seriously threatened by the end of the draft."[56] Without the draft, WRAIR lost an important recruiting tool at a time when scientists and physicians were already beginning to question a career in military medicine.

An independent review of DOD medical research programs conducted in the 1960s foresaw many of the professional and personnel problems that would contribute to the exodus of talent from military research institutes after the Vietnam War. The report identified the salary structure for senior scientists as a major source of the problem: "Fixed by statute, it cannot compete with industrial salary structures."[57] Excellent facilities and competitive salaries seduced talented scientists early in their training, but poor long-term professional opportunities within the military made it difficult to retain good scientists. In particular, "when [a scientist] reaches the level of senior investigator, a reverse process takes place. At this level, government salaries and other benefits are below those of universities and industry. Consequently the government is deprived largely of the peak productivity of a professionally mature scientist. This in turn makes it more difficult to hold the younger scientists as there are so few outstanding men left to attract them."[58] As a consequence, "during the past seven years, a staff of investigators, once recognized as one of the strongest of its kind in the world, has been literally decimated by resignations."[59]

Maurice Hilleman and James Sorrentino were among those who resigned from WRAIR for industry positions. During the postwar era, when military-industrial networks were still strong, these moves were not a zero-sum game for the military for they often reinforced working relationships with the vaccine industry. Once military-industrial collaboration weakened in the late 1970s and 1980s, however, the exodus of talent had few positive spin-offs for military research.

A growing professional divide between military officers and military scientists was also beginning to compromise research activities. Vannevar Bush believed that World War II had wrought a permanent solution to this problem. In 1970, he wrote: "The obstructionism of military systems, as it existed for a thousand years, ended with the last Great War. It is far more possible today to maintain a productive collaboration between military men on the one hand and civilian scientists and engineers on the other than it ever was before."[60] He noted that prior to this war, "Military laboratories were dominated by officers who made it utterly clear that scientists or engineers employed in these laboratories were of a lower caste of society. When contracts were issued, the conditions and objectives were rigidly controlled by officers whose understanding of science was rudimentary, to say the least."[61]

After the Vietnam War, the military suffered a relapse of creeping obstructionism and rigidity that bore an uncanny resemblance to prewar military organizations. The U.S. Army Medical R&D Command, which directed the activities of WRAIR and the U.S. Army Medical Research Institute for Infectious Diseases (USAMRIID), began to embrace an ever-growing repertoire of standard operating procedures to direct research, development, and procurement. Over time, this bureaucracy began to restrict the freedom of scientific directors to manage their own labs and relationships with collaborators. By the mid-1980s, lab directors lost much of their authority to professionals and procurement specialists.

Philip Russell, director and commandant of the WRAIR from 1979 to 1983 and commander of the U.S. Army Medical R&D Command from 1986 to 1990, witnessed many of these transformations firsthand: "I was the last director that let laboratory commanders run their own organization . . . I would let department directors decide how to use their own resources. Now research is micro-managed by the R&D Command." According to Russell, "There was a concept pushed from the DOD on down to the Department of the Army that lab scientists can't manage anything and that you need professional managers." Now, he observed, "military labs operate as contractors. They have to petition the bureaucracy to get money."[62]

This form of management from nonscientists, or what Russell affectionately called "headquarter weenies," "always results in weak programs."[63] These new managers did not appreciate the complexities of vaccine development. Their efforts to improve efficiency and cut costs also undermined WRAIR's collaborative research networks. According to Russell, "It takes the scientists to know where the cutting edges are and who is really doing the good work in the field."[64] When scientists lost managerial authority, their research programs began to suffer from disjointed missions, short-term planning, and high turnover.

Thomas Monath, former chief of virology at USAMRIID (1989–1992), felt that the military medical research bureaucracy cut both ways. "On the one hand," he explained, "there is so much scientific review, institutional review and infrastructure to help you out. They can back you up with a clinical lab; you could easily put your hands on the necessary expertise, etc. Yet the system was complicated and it was difficult to order supplies, and the bureaucracy was so large that there were a lot of junior irrelevant people to get in your way."[65]

Monath, like many other experienced vaccine research scientists during this period, eventually left the military for industry.[66] The DOD's failure to "give a man his head"—as Bush once advised—was slowly destroying the capabilities of its research organizations and causing them to bleed talent.

Military efforts to obtain an adenovirus vaccine showcase the effect of these bureaucratic changes on the gradual breakdown of the military-industrial partnerships. Adenoviruses, which cause upper respiratory disease, often infect military recruits in the first few weeks of training. Prior to the development of the vaccine, up to 80 percent of new recruits suffered from acute respiratory disease (ARD), with 20 percent requiring hospitalization in the winter months.[67]

In the early 1950s, Hilleman isolated one of the first adenovirus strains at approximately the same time that Wallace Rowe, working in Robert Huebner's lab at the National Institutes of Health (NIH), isolated several related strains.[68] Hilleman and Rowe quickly developed formalin-inactivated vaccines grown in monkey kidney cells. With the help of WRAIR's Preventive Medicine Division, they worked with military clinicians and epidemiologists to test the safety and efficacy of these vaccines.[69] By 1957, Lederle licensed WRAIR's bivalent vaccine for adenovirus types 4 and 7 and Parke-Davis licensed the NIH's trivalent vaccine for types 3, 4, and 7.[70]

Since WRAIR had already conducted in-house safety and efficacy trials, it was a relatively small step for industry to license and manufacture the vaccine. Even so, it was not easy to persuade industry to take on this project. Adenovirus-associated ARD runs rampant among new military recruits, but it does not threaten civilians to the same degree, which makes these vaccines commercially unattractive.[71] In a letter to Thomas Francis, president of the AFEB, Colonel Arthur Long, Chief of the Preventive Medicine Division, noted:

> The road to the commercial production of adenovirus vaccine appears to have been a very rough and rocky one. This came to our attention shortly after the original invitations to bid [for a second vaccine that combined adenovirus and influenza] were issued in the early summer of 1958. The various manufacturers concerned . . . demonstrated neither capability nor interest . . . there were numerous technical difficulties involved in the production of a satisfactory vaccine. It was also learned as time went by that there were some management problems involved as well since this agent could be expected to have relatively little, if any, civilian appeal.[72]

WRAIR considered producing the vaccine in-house, but soon discovered the limitations of that approach. Long observed: "At

WRAIR there is no experience in or facility for the large scale tissue culture production of vaccine. It is also realized that a great deal of time and cost would be involved in that institution reaching the current level of the commercial manufacturers' ability."[73]

With new appreciation for their dependence on industry's expertise, the Preventive Medicine Division reapplied itself to the problem of engaging the private sector. Colonel Long prevailed on the executive vice president of the Pharmaceutical Research and Manufacturers' Association (PhRMA) "to interest management in the military problem."[74] In the 1950s, responding to military needs still conferred public-relations benefits for industry. With a little extra pressure from PhRMA, "increased interest became apparent."[75] Colonel Long reported that now "at least two and possibly other companies [were] interested in bidding on a purchase proposal."[76]

Once the military secured a regular supply of the vaccine, rates of ARD went down considerably. The Surgeon General's Office (SGO) noted in 1960 that hospital admission rates were down one-third from the year prior, stating, "We hope this is due to adenovirus vaccine but we do not want to give out any advance publicity until we are certain."[77] Just as the army was making headway against this disease, they began to experience supply problems with the manufacturer. The SGO stated, "Our stock of adenovirus is now depleted. We have indicated to the manufacturers that we are most unhappy with the supply of this vaccine."[78]

The situation worsened when researchers determined that the monkey kidney cells used to grow the virus were contaminated with SV-40, a cancer-causing virus. Safety concerns prompted the BOB to terminate the Lederle and Parke-Davis licenses in 1963 and the military ran through their existing stocks within the year. The SGO reported, "There has been no adenovirus vaccine available for routine use since early 1964. Commercial producers in coordination with U.S. Army Medical Research and Development Command have expended considerable effort to find the solution to production problems."[79]

Scientists at WRAIR (Edward Buescher) and the NIH (Robert Channock) worked quickly to reformulate a safer adenovirus vaccine. They developed a live oral adenovirus vaccine grown in human embryonic kidney cells. When swallowed, it infected recruits by an alternate route and caused them to develop immunity to the more serious respiratory form of the disease.

In contrast to Hilleman's and Rowe's efforts, which had been competitive, this second collaboration was more congenial. According to Russell, "The two [Buescher and Channock] were close friends, both had been fellows in Albert Sabin's laboratory and both had served at the army's 406 General Medical Laboratory in Japan conducting virus research. Between them, they overcame the institutional rivalries of the past and, in collaboration with Ben Rubin at Wyeth Laboratories, provided the leadership that resulted in the very successful development of the type 4 and type 7 oral adenovirus vaccines."[80]

NIH and WRAIR investigators tweaked the vaccine to ensure safety and efficacy and worked closely with Wyeth to develop processes to grow the virus, lyophilize the vaccine, and compress it into a tablet that would maintain stability. Wyeth began producing the vaccine for routine use in the military under Investigational New Drug (IND) provisions in 1971. According to Russell, once the military initiated routine vaccinations, "the large hospital wards at recruit camps formerly used to treat ARD cases were closed."[81] According to a study from the 1970s, this vaccine saved the U.S. Army $7.5 million in medical costs after two years of routine use, whereas the vaccine itself incurred only $4.8 million in development and testing costs.[82] The impressive safety record compiled from recruit vaccinations over the years allowed Wyeth to obtain a license for the vaccine in 1980.[83]

Despite the clear value of the adenovirus vaccine, it received insufficient support from DOD procurement officers. When military officers asked Wyeth to manufacture their new live-oral adenovirus vaccine, Rubin countered that the proposed contract was unrealistic. He explained that more research, requiring long-term, consistent funding, was needed before the live-oral vaccine could be scaled up for production, and his lab could not perform this work without compensation.[84] Increasingly inflexible military funding practices made it difficult for them to accommodate Wyeth. One AFEB member explained, "Because it takes so much time to get from there to here, this is not compatible with military funding."[85] Rubin concluded the meeting with a warning: "Let's make it perfectly clear. Wyeth is going to get out of the business, and furthermore, most of the other companies that are producing vaccines are going to get out of the business too, and it's probably going to be a lot sooner than you think."[86]

In 1984 Wyeth asked the DOD if they would assist with the cost of renovating the adenovirus vaccine facility—estimated at $5 million. The DOD refused. According to Monath, "At the time, they probably just didn't have the extra money needed. There was no way to advance the funds and so they couldn't negotiate."[87] He explained that Wyeth threw up their hands and countered, "We're not making any money from it. It is just a service."[88] Wyeth ceased producing the vaccine, and by 1996 the military, once again, stopped vaccinating troops. Monath explained, "We're now having epidemics in recruits of adenovirus types 4 and 7 again, as well type 3. And there has been no manufacturer. The Army has been unable to find a manufacturer because the market is so small. So it's a real embarrassment. Now it will cost them $50 million at least to reconstitute this thing. You could have done it for $5 million 5 years ago—just absolute crass stupidity."[89]

Some preventive medicine officials were shocked when they had to terminate their adenovirus vaccination program due to a mere supply problem. They should have reserved their amazement for the fact that they had been able to engage industry in a low-margin, limited-use vaccine contract for as long as they had. Some older AFEB members were less surprised by this outcome.

Several decades earlier, the AFEB had explored the option of building a government-owned, company-operated (GOCO) vaccine development facility when it was struggling find an industrial partner to develop the meningitis vaccine. In 1972, Colonel Robert Cutting explained that "as a result of Squibb's withdrawal from the meningitis vaccine, and after a great deal of hand-wringing in anguish, an internal emergency committee was created at the DOD level, Dr. Wilbur's office, for the purpose of addressing this very specific question."[90]

The Commission on Acute Respiratory Diseases recommended that "the AFEB in collaboration with the DOD and the PHS, investigate the feasibility that the U.S. Government develop a facility whose major function would be the research and development of vaccines. The possibility of the production of certain vaccines might also be considered for those preparations that are not commercially profitable."[91] Floyd Denny, a member of the Commission on Acute Respiratory Diseases emphasized, "There is indeed a huge problem. This applies not only to adenovirus vaccines but other vaccines as well and I think this needs to be addressed very, very emphatically now."[92]

The SGO had already had some success with this model. In 1960, they built a contract manufacturing facility to develop limited-use vaccines at the National Drug Company in Swiftwater, Pennsylvania. They eventually donated the facility to the Salk Institute, which operated as a nonprofit on the Swiftwater campus until 1998.[93] The Salk Institute continued to develop pilot lots of limited-use vaccines for WRAIR for another twenty years. Between 1960 and 1998, this facility filed INDs for vaccines against Junin virus, chikungunya virus, eastern equine encephalitis, western equine encephalitis, Venezuelan equine encephalitis, Q fever, Rift Valley fever, and tularemia. Most INDs were filed in the 1960s and 1970s, and manufacturing ceased for most of these limited-use vaccines in the 1970s and 1980s (see Table 1.2, Chapter 1).

Encouraged by their experience with the Swiftwater facility, heads of the preventive medicine divisions of all three services wrote letters in support of building a GOCO to develop and manufacture a wider range of vaccines for the military. In a move that would become increasingly uncommon, however, industry stepped in and saved the day. According to Colonel Cutting, "People were dancing around, trying to come to grips with this problem when all of a sudden Merck decided that they would make the meningitis vaccine."[94] Merck agreed to make the vaccine, not for business reasons, but because Hilleman's former colleagues at WRAIR were able to prevail on his sense of loyalty and obligation.

Russell suspected that industry's former willingness to take on military vaccine projects stemmed less from than an abstract sense of duty and had more to do with the personal relationships between scientists at WRAIR and in industry.[95] As professional managers began to assume more responsibility for administrative duties that traditionally fell to the scientists themselves, such as contract negotiations, long-standing working relationships and personal ties were broken. Under these circumstances, both sides often chose to cancel existing contracts and to reject new ones.

The "professionalization" of management within the DOD was a response to a bipartisan push to import private-sector practices toward the end of the Cold War. Reforms consisted of downsizing, centralizing in-house activities, and outsourcing the rest. As commercial innovation began to outpace military contributions in electronics and computing, there seemed to be little room for argument. In a comparative study of new warships and M14 rifles, Aaron Friedberg demonstrated what most already believed: that "private firms consistently gave better value for the tax payer's dollar."[96] However salutary these techniques may have been for national defense as a whole, it crippled military vaccine research and development activities.

In a bout of centralization in the 1990s, the DOD placed responsibility for biodefense vaccine procurement in the Joint Development Office (JDO). According to Russell, this was a "disastrous policy decision," for it isolated the research process from the medical departments and from anyone who understood the vaccine industry.[97] The crowning blow to WRAIR came in 1994, when the DOD moved the operational capabilities of the Preventive Medicine Division from WRAIR to Aberdeen Proving Ground in Maryland. This reorganization was part of a larger effort to centralize preventive medicine activities at WRAIR, the SGO and the Environmental Hygiene Laboratory into the Center for Health Promotion and Preventive Medicine (CHPPM) at Aberdeen. According to Patrick Kelley, former director of Global Emerging Infections Surveillance at WRAIR (DOD-GEIS), this move made sense from "an organizational command and control point of view" because it consolidated operational public health activities for the army, but the quality of research may have suffered as a result.[98] Joel Gaydos, who directed the clinical preventive medicine program at the newly formed CHPPM, thought that this move offered a tremendous opportunity to strengthen epidemiologic capabilities. At CHPPM, WRAIR's epidemiological consultation teams (EPICON), which conduct fieldwork such as disease identification and clinical trials, were given a well-defined role, an adequate budget and staff, and a policy function that afforded long-term strategic planning. In hindsight, however, he conceded that the reorganization did not live up to its promise.

CHPPM became bogged down in routine occupational health and environmental medicine service requirements and strategic planning for vaccines drifted into the background. Responsibility for the strategic planning and public health assessment of military vaccine needs became disaggregated and important problems fell through the cracks. According to Gaydos, preventive medicine and infectious diseases physicians tried unsuccessfully for years to simply identify the office or organization responsible for addressing the most important military vaccine problem at the time, the loss of the adenovirus vaccine.[99]

Transferring EPICON teams to CHPPM "was a bad move," according to Russell, "because it removed the field capability that the preventive medicine division had from the intellectual interaction they had with the research folks at WRAIR. To separate the two was monumental stupidity."[100]

Prior to the move, WRAIR possessed a fully integrated set of epidemiological, clinical, and lab capabilities. These groups worked together under WRAIR's Preventive Medicine Division to capture lead-user insights and to integrate field and lab research. Russell recalled that he could put together teams from each division to "solve almost any problem."[101] Unlike many of the vaccine candidates that came out of university labs, or even the NIH, WRAIR was able to hand industry candidates that had been fully developed and tested through their clinical trial infrastructure at home and abroad. In this way, the military was able to assume much of the cost and risk of vaccine development for industry.

When WRAIR lost the operational arm of the Preventive Medicine Division to CHPPM (and many of their international labs), they were not able to generate and test new vaccine candidates with the same rigor. This created a lacuna in the development cycle for vaccines, especially biodefense and global health vaccines, which had enjoyed military sponsorship in the past.

The DOD revisited the GOCO issue in 1993 when it failed to procure sufficient amounts of anthrax vaccine and botulinum antitoxin prior to the first Gulf War. A triservice task force concluded that a GOCO would not only generate a stable supply of medical countermeasures (MCMs) but could provide much-needed surge capacity in a crisis.[102]

To alleviate supply constraints in the short term, the FDA passed an interim rule waiving informed consent requirements for the use of investigational new drugs and vaccines (INDs) to protect the troops.[103] This decision was later widely criticized as INDs such as pyridostigmine bromide and botulinum toxoid fell under suspicion for causing the Gulf War syndrome. Medical and political fallout from the Gulf War syndrome, coupled with the impracticality of adhering to informed consent regulations in a rapid mobilization scenario, prompted military officials to require FDA approval for all MCMs.

This new requirement prompted the DOD to create the Joint Vaccine Acquisition Program (JVAP) and hire a lead system integrator (LSI), DynPort, to farm out military contracts to procure vaccines in the marketplace. When DynPort failed to attract good industry partners, a panel of military and industry experts conducted another investigation of the DOD procurement strategy in 2001. Their report (the Top Report) questioned the longstanding predilection for market-based solutions to military research and development needs and, once again, proposed a compromise between "out-sourced" and "in-house" solutions. The report acknowledged that "large, well-established pharmaceutical industry manufacturers [were] unlikely to reverse their decades-long trend of relatively inconsequential support for DOD vaccine production requirements."[104] It also noted that "the scope and complexity of the DOD biological warfare defense vaccine requirements were too great for either the DOD or the pharmaceutical industry to accomplish alone."[105] To overcome these obstacles, the panel proposed a GOCO vaccine research, development, and manufacturing facility to develop, license, and manufacture vaccines for biological warfare agents and endemic disease threats of military significance.

The DOD ultimately shelved the Top Report in favor of continued efforts to procure vaccines through commercial channels. While government officials and legislators objected to the cost ($1.56 billion to build, maintain, and operate the facility over a twenty-five-year period), industry objected to the proposal on principle. Industry had thus far been unwilling to develop the handful of limited-use vaccines under consideration, even with a host of government-sponsored cost- and risk-sharing options. Nonetheless, they feared that the project could encroach on viable markets. Wayne Pisano, executive vice president of Aventis Pasteur, elaborated on industry's concerns in mildly apocalyptic terms: "If the government enters the vaccine manufacturing business, industry may ultimately be unable to compete with a subsidized operation. The result is, potentially, the withdrawal of private manufactures from the U.S. market and the loss of innovation and timely introduction of new products into the U.S."[106]

This proposal was also unpopular with many policy makers. The federal government, they argued, should not be in the business of producing anything. Even though NASA had put a man on the moon as a result of in-house engineering and manufacturing, any mention of the agency was less likely to evoke these historic successes and more likely to conjure memories of faulty O-rings (the 1986 Challenger disaster) and faulty thermal shields (the 2003 Columbia disaster). As the Dow Jones Industrial Average continued to climb through the 1990s and early 2000s, military planners and legislators were increasingly willing to place their faith in the private sector.

Policy makers and acquisition managers were drawn to market-based governance schemes in which they outsourced a growing array of critical functions to reduce the role of the government as the provider of goods and many services. The DOD began to outsource an ever-greater portion of research. According to one study, navy labs outsourced 50 percent of their workload in 2000, up from 26 percent in 1969. Army labs outsourced 65 percent, up from 38 percent.[107] By 2009, even NASA—one of the most visible government agencies with production capabilities—agreed to start contracting with private companies to design, build, and service their own rockets and spacecraft.

The DOD soon began to outsource the task of managing extramural research and development as well by hiring LSIs to manage large federal acquisition programs. LSIs are primary contractors that manage a collection of secondary contractors for the armed forces. After critical MCM shortfalls and safety concerns bedeviled biodefense in the first Gulf War, the JDO established JVAP in 1994 to bring experimental biodefense vaccines to licensure. JVAP turned around and awarded $747 million to DynPort, a British-American company, to contract private firms to develop and test seventeen biodefense vaccines.

Once the DOD established JVAP and hired DynPort to farm out military contracts for biodefense vaccines, it decommissioned the Salk Institute in 1998. While the objective was to upgrade military biodefense vaccine manufacturing capabilities, this move eroded yet another traditional source of hands-on vaccine development knowledge. Even when the Salk Institute failed to perform on a contract, it still developed skills, maintained working relationships with WRAIR scientists, and retained an institutional memory of lessons learned. Today, however, when DynPort fails, its failure is complete. DynPort terminates the contract and former collaborators may never work together again. Results are not published and lessons are not retained.[108]

Hiring an LSI to farm out JVAP contracts also compounded research and development governance problems by disaggregating the process further. According to Robert House, DynPort's chief scientific officer, "We are responsible for designing most of the experiments, as well as interpreting the results. In this respect, you might say we are conducting our research via remote control."[109] When contractors do not conduct research of their own, it can be harder for them to make informed decisions about the quality and progress of subcontracted work. This has become doubly true for the Pentagon, which must now work through the additional layer of bureaucracy that LSIs impose. Over time, the Pentagon leadership has become so far removed from the research and development that underlies its acquisition options that it "is losing its ability to tell the difference between sound and unsound decisions on innovative technology and is outsourcing key decision-making as well."[110]

New cadres of professional acquisition managers at the Pentagon did not appreciate the time, cost, and complexity of vaccine development, in particular. Monath observed that the army began to suffer from the conception that "the process of product development was a mild modification of how the army develops a new tank . . . getting to that last step with products was exceedingly difficult because the tank mentality doesn't apply to pharmaceuticals very well." He concluded: "So there are a lot of people who get involved at the end stage here who really don't know what they are doing."[111]

The Pentagon was seduced by another popular management technique: breaking down development tasks into pieces and contracting them out to the lowest bidder. This modular approach presumes a near-perfect understanding of a process that can be prespecified and made to order. Unlike some military hardware, which can be assembled to specification through an interchangeable network of subcontractors, vaccine-relevant knowledge can be difficult to quantify and transfer. The vaccine development process is fickle and often resists accurate cost projections, timing, and yield estimates. The knowledge required to successfully grow and manipulate biological material is contextual and often tacit.

Vaccines are not tanks. They are not easily reduced to a blueprint that can be passed down a chain of arm's-length transactions. Yet this is what happens when LSIs outsource development projects. While our understanding of biological processes may one day allow a more rational approach to vaccine design, modularity—and the horizontal governance structures that it permits—is an inappropriate model for many aspects of vaccine development.

In addition to disaggregating the research process, outsourcing vaccine research and development exacerbates preexisting market incentive problems. When the DOD hired DynPort to manage JVAP, they simply shifted acquisition responsibilities further into the marketplace, without removing any of the original barriers to market-based solutions. These contracts remain unattractive to large, experienced biopharmaceutical companies regardless of who administers them. Instead, contracts go to smaller, less experienced companies that do not carry the same opportunity costs as larger firms. As these companies struggle to perform the work of a large integrated firm, contracts inevitably have to be renegotiated, extended to a wider array of companies, or terminated. This process elevates the time, cost, and failure rate of these projects. To date, this program has failed to license a single new vaccine.

~ AS MILITARY MEDICAL research labs began to lose talent, funding, and prestige during the post-Vietnam period, the NIH began to gain in all three areas. NIH support for academic research continued to grow in the midst of the government wide budget cuts of the 1970s, initiating a trend that persisted throughout the 1980s and 1990s.[112]

In 1980, the NIH's National Institute for Allergies and Infectious Diseases (NIAID) designated itself as a center for "next-generation" vaccine research and initiated a Program for Accelerated Development of New Vaccines. According to one report, "The incentive for an expanded effort lay in new knowledge and processes emerging from recombinant DNA and hybridoma technologies, and in the better understanding of the workings of the immune system."[113] The idea was to select high-priority vaccines, leverage new recombinant technologies, and target research objectives to accelerate their development. By 1986, however, the program had contributed to the development of only one vaccine containing antigens derived from recombinant DNA expressed in yeast cells, the hepatitis B vaccine.

William Jordan, director of the Microbiology and Infectious Diseases Program at NIAID, acknowledged that the development of genetically engineered vaccines was not proceeding as quickly as expected.[114] Jordan catalogued a number of scientific setbacks: synthetic oligopeptides were not as antigenic as hoped, it was difficult to demonstrate the safety of antigens produced in continuous mammalian cell lines, and polysaccharides had to be coupled with proteins to be immunogenic in young children.[115] The problems facing NIAID's vaccine development programs were, from Jordan's point of view, entirely scientific in nature.

Military and industrial vaccine research scientists familiar with NIAID's vaccine development program disagreed. They identified a number of organizational and managerial characteristics that they felt were counterproductive for vaccine development. Colin MacLeod, former AFEB president, compared the organization and efficacy of AFEB commissions against the value of NIH study sections. He explained that AFEB commissions were made up of experts in a well-defined disease-oriented area who devised targeted research programs in collaboration with medical departments of the military services. NIH study sections, on the other hand, "are not goal-oriented and usually do not have any responsibility for the development of a program . . . They are passive, judicious bodies whose function is to review applications sent to them."[116]

Hilleman also took a dim view of NIH contract research programs: "These are little cottage industries, aren't they? And they are different disciplines, different programs. Now, in these little cottage industries, you can't bring central focus . . . All of these independent diddleworks never get coordinated to provide solutions."[117]

The NIH does not offer an oversight body to integrate findings or to bring research projects in line with a cohesive vaccine development plan. Rather, they encourage scientists to pursue individual lines of inquiry simultaneously. This independent investigator-initiated, peer-reviewed grant system generates scientific knowledge and publications, but it falters when the government requests tangible results, as it did when the organization was called upon to produce a vaccine to prevent AIDS.

In 1984, Robert Gallo announced that he, along with his colleagues at the National Cancer Institute (NCI), had identified the virus that causes AIDS. In a few short years, NIAID led the search for an AIDS vaccine, under Antony Fauci, with a budget of $261 million. By 1999, the budget would grow to $1.8 billion, but no vaccine was in sight.[118]

HIV viruses present unique challenges that would have thwarted vaccine research scientists under the best of circumstances. Some contend, however, that scientists would be much closer to a vaccine if the NIH had organized research differently. Hilleman argued that the NIH's "fractured organization, centered on individual investigator-initiated RO1 grants, allowed it for too long a time to pursue a subunit envelope vaccine against AIDS as a continuing paradigm to the highly effective subunit hepatitis B vaccine of 1981 and 1986 that can prevent infection by antibodies alone."[119] Had there been more top-down oversight, he argued, efforts would have been redirected towards an examination of cell-mediated immune responses much sooner.

In some respects, NIAID was a victim of its own success. Vannevar Bush foresaw the dangers of excess funds in the 1950s, just before large appropriations for biomedical research became the fashion. He was concerned that excessive funding would permit mediocre research and encourage overspecialization, leading to the construction of a scientific "Tower of Babel" that would thwart efforts to integrate findings and coordinate research programs.[120] While Bush was concerned for the future of integrative research efforts in general, this issue was particularly relevant to vaccine research scientists who need to integrate findings from a diverse number of fields and specialties.

Hilleman suggested that additional money and research were unlikely to yield a vaccine but that a reorganization of the research program itself could. He blamed the NIH grant system for encouraging scientists to pursue "what is the more doable at the expense of what is more urgently needed (. . .) the total of basic knowledge in AIDS, particularly its immunology and molecular biology, is overabundant and is highly complex . . . the more fundamental problem for achieving an AIDS vaccine may lie with organizational structure and management than in attaining the needed science itself. Pursuit of an AIDS vaccine could be benefited greatly by a determination of what areas of missing information are important and how these needs could be organized into defined areas of emphasis for investigation by separate team efforts."[121]

Hilleman's call for a more top-down approach, while abhorrent to some scientists, began to hold sway with others. In 1996, the Levine Committee (a committee of one hundred scientists and activists, which had a "list of participants that read like a Who's Who in academic AIDS research" issued a report on the state of vaccine research at NIH.[122] The report concluded unequivocally that "HIV vaccine research and development is in crisis" and that "the entire AIDS vaccine research effort at NIH should be restructured."[123]

The committee argued that "the NIH must be prepared to go beyond its traditional role, for the discovery and development of a vaccine demands more than just the acquisition of fundamental knowledge; it requires that the information be applied and resultant vaccine strategies appropriately evaluated."[124] To accomplish this, the Levine Committee advocated OSRD-style top-down coordination and integration of research objectives. In particular, they recommended an interdisciplinary study section for vaccine research that would set priorities and direct research objectives. It was, however, difficult to implement integrated programs in the context of the NIH's strong bottom-up culture and the Clinton Administration decided to construct a separate Vaccine Research Center to coordinate these efforts.[125]

Industry scrambled to profit from the mountains of data coming out of the NIH and academia as well. Recombinant techniques promised new opportunities to design vaccines with greater specificity and predictability. The prospect of being able to engineer an antigen apart from its pathogen represented a significant opportunity to develop vaccines with fewer side effects. This was particularly attractive for vaccine producers operating in the litigious climate of the 1980s.

The explosion of technological opportunities in molecular biology and genetic engineering made it progressively more difficult for any one firm to assemble the optimal array of in-house development expertise. The 1980 passage of the Bayh-Dole Act exacerbated efforts to integrate insights generated by the biotechnology revolution. This legislation allowed universities, small businesses, and nonprofits to obtain intellectual property (IP) rights to their inventions. Gone were the days of "free exchange" that accelerated mid-century vaccine development. Increasingly, firms had to navigate multiple patents to assemble the necessary knowledge and tools required to develop a vaccine. Upstream discoveries and research tools were underused because a tangle of IP protections made them inaccessible and expensive.[126]

Biotechnology entrepreneur Robert Carlson notes that "the cost of participating in the market is not just the capital and the labor required to produce a new object for sale; it includes the cost of protecting the resulting intellectual property." He further observed, "In my case, the entire capital cost of product development was substantially smaller than the initial transaction costs of writing and filing a patent."[127]

Commercial vaccine developers responded to the rising complexity, cost, and risk of development much in the same way that military labs did, by expanding their reliance on outsourcing and technological cooperation agreements. While some firms invested in in-house capabilities, others formed contractual arrangements with smaller biotechnology start-ups and academic and government research institutions.[128]

Many industry observers regard this trend towards decentralization as an unalloyed boon to productivity. Outsourcing and technology cooperation agreements, they argue, help firms cope with the rapidly changing landscape for technology development.[129] Subcontracting is attractive because it allows firms to specialize in areas where they have a competitive advantage and to outsource in areas where they do not.[130] Technology cooperation agreements, they maintain, allow firms to seize new opportunities while managing their risks and costs.[131]

Luigi Orsenigo argues that horizontal arrangements are not only an efficient solution to the growing cost and complexity of production, they are a valuable source of innovation.[132] In theory, vaccine companies have much to gain from close working relationships with partners that share heterogeneous, yet complementary, research and development capabilities. These dense networks of collaborative relationships, he argues, serve as "organizational devices for the coordination of heterogeneous learning processes by agents characterized by different skills, competencies, access to information and assets."[133]

Early evidence from industry seemed to support these ideas. Merck leaned heavily on outside networks to develop a recombinant hepatitis B vaccine in the 1980s. This vaccine was the product of a four-way collaboration between Merck, two university laboratories (one headed by William Rutter at the University of California and the other by Benjamin Hall at the University of Washington), and Chiron, a California-based biotechnology company.[134] Together they developed a system to express hepatitis B surface antigens in yeast, resulting in the first safe and effective recombinant vaccine, licensed in 1986.

The success of this collaborative model was, however, hard to replicate. Apart from SmithKline Beecham's version of a recombinant hepatitis B vaccine (introduced in 1989), CBER has licensed only two other recombinant vaccines: a Lyme disease vaccine (licensed in 1998, withdrawn in 2002) and the human papilloma virus vaccine (licensed in 2006).

Champions of highly decentralized governance structures underestimate the extent to which these "dense networks" carry high transaction costs and often fail to deliver the advertised knowledge spillovers. When the information required for technological problem solving is sticky, as it is in vaccine development, these collaborative ventures can be less effective.

The research habits of NIH-trained scientists complicated efforts to maintain broad, interdisciplinary and translational research programs. By 1982, the NIH was the unquestioned leader in the field of molecular biology, having spent an estimated $380 million on research in this area.[135] As industry hastened to acquire biotechnology capabilities, their interest was further deflected from the military and toward the NIH as a source for new hires and collaborative partners. This had a profound impact on the composition, capabilities, and practices of vaccine research and development communities.

Asked why Merck had not produced more than two new vaccines in the 1990s (hepatitis A and varicella), Hilleman replied, "I believe it is because Hilleman is not there anymore."[136] Beneath the self-effacing humor, he made a valid point. Hilleman brought WRAIR-style bench-to-batch scientific management to Merck. His crossover teams worked together to transition vaccine candidates from test tubes, to fermentation tanks, to field tests. After Vietnam, however, as WRAIR lost talented researchers, funding, and operational capabilities in the field, this unique talent pool of "science integrators" began to disappear.

By contrast, the NIH bred specialists. When Hilleman retired in

1984, Roy Vagelos hired his replacement, Edward Scolnick, from the genetics division of the NCI. Under Scolnick's leadership, Merck's vaccine division reverted to a more bureaucratized form of scientific management with greater separation by specialization.

As integrated research and development practices fell by the wayside in government and industrial labs alike, the number of new or improved vaccine licensures declined (Figure 2.1). There was a general sense among viral vaccine developers that the technological opportunities afforded by cell culture techniques had been largely exploited. According to Hilleman, "Much of the pioneering for vaccines had been completed by 1984 . . . this was true for the entire industry."[137]

Dwindling technological opportunities cannot, however, fully account for falling rates of innovation during the latter half of the century. Just as what some industry observers have called the "virology cycle" was winding down, a new cycle based on molecular biology techniques was winding up.[138] An entirely new industry had grown up around biotechnology to support the development and application of these tools. Nonetheless, access to new technologies and firm capabilities failed to yield the same high rates of innovation witnessed in the 1940s through the 1960s.

There was also a sense that economic opportunities for vaccine development had diminished as liability and regulatory costs rose and vaccine producers exited the industry in droves. Industry consolidation, however, left the remaining firms with virtual monopolies over individual vaccine product lines. In many cases, firms raised their prices without having to make additional investments in their vaccine products. Profits rose, but new profits did not always translate into new vaccine products.

To understand the downward shift in vaccine innovation in the 1970s, it is important to look beyond these traditional measures to consider broader changes in the landscape for vaccine development. These changes range from public relations fiascos, to a shift in the balance of power from the DOD to the NIH, to the rise of stronger intellectual property regimes, and the collapse of open exchange collaborative research networks. These changes disrupted working relationships within and between military and industrial labs. As military-industrial collaborative networks began to dissolve, industry lost access to a valuable development partner and had to work harder to translate NIH and university-based research into licensed vaccines.

A host of political, institutional, strategic, and historical factors have sustained this trend. Reintroducing integrated research practices would require a fundamental shift in scientific training opportunities and professional incentive structures for vaccine development. Whole-scale change of this nature would have to overcome opposition from scientists, politicians, and industrialists who are invested in the current model. Just as World War II restructured the political economy for vaccine development, one might expect that another national emergency could hasten this change. The terrorist attacks of 2001 presented one such event, creating both greater challenges and opportunities for the future of vaccine development.

6 Biodefense in the Twenty-First Century

BIODEFENSE INITIATIVES BOOMED after the airliner and anthrax attacks of 2001. While the overall budget to fight terrorism doubled after 9/11, the bioterrorism budget quadrupled.[1] New vaccines "to fight anthrax and other diseases" were featured in the 2002 State of the Union address, and the Bush administration championed a biodefense vaccine development campaign that seemed destined to rival the successes of World War II programs.[2]

Yet, a decade after these attacks, the United States has licensed only one new biodefense vaccine—the ACAM2000 for smallpox.[3] Notably, this vaccine was developed before post-9/11 vaccine development programs took effect. Why are twenty-first-century programs producing vaccines at such a glacial pace compared to earlier efforts undertaken during World War II?

World War II programs succeeded where current programs have failed on one critical dimension: the ability to consolidate and apply vaccine-relevant knowledge and skills. Then, as now, this ability hinged on close public-private collaboration because critical late-stage development skills and manufacturing capacity resides in the private sector.

Higher levels of cooperation were easier to achieve in the 1940s in part because public and private interests were united by a severe external threat. The urgency of war provided the fuel for these efforts and the OSRD and SGO provided the focus, concentrating funding streams and attention spans on the immediate problem of getting new vaccines to the troops.

Subsequent wars (Vietnam, Iraq, the "war on terror") invoked threats and responses that were less acute and more disputed. The external threat was not as high and industrial, academic, military, and congressional interests were not aligned. Without strong, targeted programs backed by all four groups, vaccine development proceeded in a relatively uncoordinated fashion.

Personal histories mattered too. Pathogenic threats were particularly salient to the World War II generation, which had survived childhood and World War I without the benefit of pediatric vaccines or antibiotics. Many war planners experienced the ravages of the 1918 pandemic firsthand and had deep respect for the humanitarian and military value of vaccines, whereas postwar generations have been more hesitant to initiate vaccine campaigns. The legal fallout of the 1976 swine flu affair and unanswered questions about the Gulf War syndrome linger in the public memory today, and concerns about the side effects of a vaccine often outweigh concerns about the disease itself.[4]

The 2001 terrorist attacks suspended this hesitant mood for a short time. Fear and uncertainty about new biological attacks at home encouraged both industry and the government to break out of established routines to find new ways to cooperate on vaccine development. After decades of fruitless efforts to develop limited-use vaccines, federal agencies were suddenly besieged with offers to help.[5]

The pharmaceutical industry began to use patriotic rhetoric that was reminiscent of World War II–era proclamations. Alan Holmer, president of the Pharmaceutical Research and Manufacturers' Association (PhRMA), claimed that there was an "overwhelming desire" on the part of the pharmaceutical industry to help the government because "We're Americans first."[6] Alluding to the extraordinary efforts of pharmaceutical companies during World War II to develop antibiotics and vaccines, he boasted that "American pharmaceutical companies have been there in the past for the country in times of national crisis."[7] In a press conference with Health and Human Services (HHS) Secretary Tommy Thompson, Holmer announced, "We are offering the government all of our resources . . . We are united with our government and the American people. We are not going to let the terrorists succeed."[8]

In the fall of 2001, industry support was not merely rhetorical. Pharmaceutical executives rushed to Washington with biodefense proposals. Gail Cassell, vice president at Eli Lilly and Company, was concerned that attacks with other agents would follow. She "tore through paperwork that normally would have taken months, put samples of the drugs on a plane and flew them to government laboratories in the Washington area to be tested against smallpox."[9] Should one of Lilly's drugs prove effective, Cassell promised to initiate a crash development program, stating, "We are absolutely willing to do that. I would emphasize that we would do it for the good of the country, not for the good of Lilly."[10]

Large pharmaceutical manufacturers began to respond to Department of Defense (DOD) and HHS requests to fill contracts for smallpox and anthrax vaccines. Just weeks after the attacks, a host of companies—including Merck, GlaxoSmithKline, American Home Products, and Baxter International—submitted formal proposals to produce a smallpox vaccine.[11] In another move reminiscent of World War II, pharmaceutical companies offered to lend their own scientists to federal laboratories to work on new vaccines and therapeutics in what they described as a "gift to the nation."[12]

Government officials were equally responsive. Carol Heilman, division director at NIH, told reporters in November, "This has been a shock to so many people . . . People aren't sleeping anymore. Everybody is working as much as they possibly can. Bureaucracy is not a word that's acceptable anymore."[13] HHS processed these proposals in record time. Within a month, they announced that Baxter International, in cooperation with Acambis, a British-American biotechnology company, had won a $428 million contract to produce 155 million doses of smallpox vaccine.

Much like the early days of World War II, companies did not compete for the smallpox contract with the expectation of large financial rewards. An analyst at Morgan Stanley Dean Witter warned investors that it would be difficult for Baxter to turn a profit: "It's a competitive business, and in this sort of bidding situation, there's no way you can make money."[14]

In another parallel, genuine motives to serve the public were intermingled with opportunistic ones. Baxter had been casting about for a good public relations opportunity after it had been found liable for the deaths of over fifty individuals undergoing treatment with its dialysis machines.[15] Firms such as Merck and GlaxoSmithKline had also been losing public-relations battles in which they were accused of price gouging, thwarting generic manufacturers with patent extensions, and failing to provide affordable AIDS drugs to poor countries. Offering to assist the national biodefense effort gave these companies an opportunity to burnish their public image.

Industry's response cannot be wholly written off as a public relations ploy. The commercial appeal of developing a biodefense vaccine for the U.S. government was no stronger in 2001 than it had been in 1940. Yet over three hundred companies reported that they could develop technologies to defend against bioterrorism.[16] The United States was experiencing a wave of threat-induced industrial patriotism similar to the one witnessed on the eve of World War II. This time, however, the U.S. government failed to harness this sense of urgency into effective development programs.

During World War II, the OSRD and the Committee on Medical Research (CMR) acted quickly to convert patriotism into solid commitments by drawing top talent from industry and academia into a product development program. Vannevar Bush and Alfred Newton Richards were widely respected in industry and academia. They lent visibility and credibility to the relatively novel project of accepting federal contracts to conduct research and development. By dint of their professional reputations, political acumen, and budgetary power, they created a central clearinghouse for wartime research and development projects. In short order, the best firms and universities knew where to bring their proposals.

By contrast, no single organization or administrator emerged in the aftermath of 9/11 to corral research and development initiatives.

Early success with the smallpox vaccine contract soon proved to be an exception to the rule. Enthusiasm turned to frustration as vaccine developers struggled to determine who was in charge. Members of the industrial and academic community testified that they had workable proposals but that they did not know where to send their ideas or to whom they should talk. Worse still, they noted that the agencies themselves often did not know where to direct these calls.[17]

While the majority of medical biodefense research and development is concentrated within HHS and the DOD, these departments mirror the overall pattern of pluralism and decentralization that characterizes the national biodefense effort. The National Institute of Allergy and Infectious Diseases (NIAID) sponsors over fifty independent biodefense initiatives through intramural and extramural programs.[18] The NIAID is able to support multiple projects in parallel, but it is not designed to coordinate these projects across disciplines. DOD biodefense programs are scattered over a broader array of intramural and extramural programs administered through the Walter Reed Army Institute of Research (WRAIR), the U.S. Army Medical Research Institute for Infectious Diseases (USAMRIID), the Defense Threat Reduction Agency (DTRA), and the Defense Advanced Research Projects Agency (DARPA). In addition to these programs, the CDC, the Department of Energy (DOE), the Environmental Protection Agency (EPA), the Food and Drug Administration (FDA), and the Department of Homeland Security (DHS) also own pieces of the biodefense research agenda.

Biodefense efforts in each of these departments and agencies have proceeded through congressional initiatives that reflect a wide range of constituent interests and agency missions rather than a nationally coordinated program. Una Ryan, CEO of Avant Immunotherapeutics, which licensed a prototype recombinant anthrax vaccine to the DOD, argued that, at a minimum, "there needs to be a clearinghouse for information that would let me know exactly which government agencies, offices, and labs are responsible for research, development, procurement, and policy relevant to my products."[19]

Other Bush administration missteps dampened industry's enthusiasm further. In the midst of the anthrax attacks, HHS Secretary

Tommy Thompson decided to stockpile 1.2 billion doses of the antibiotic Cipro. Bayer offered to sell the drug to the government at a discounted price, but Thompson demanded the antibiotic at approximately half of their discounted offer. When Bayer resisted the demand, Thompson threatened to break their patent to procure a cheaper generic version of the drug. Ultimately, Bayer came down further on their price to avoid a patent-breaking debacle.[20]

Although Tommy Thompson got the price he wanted, he lost an opportunity to form critical alliances with industry. HHS sent a clear signal that intellectual property rights could not be guaranteed in the event of an emergency. Meanwhile, prospective industry partners asked themselves, "When else might stockpiles of biodefense vaccines and therapeutics be used if not in an emergency?"

A second incident, over a year later, alerted potential industry and academic partners to the liability of working with select pathogens that fell under new biosecurity regulations promulgated under the U.S.A. Patriot Act of 2001 and the Public Health Security and Bioterrorism Preparedness Act of 2002.[21] In January of 2003, Thomas Butler, chief of Infectious Diseases at Texas Tech University Medical School, ran afoul of these regulations when he failed to document the destruction of thirty vials of plague. Unable to account for the vials, Butler suggested that they might have been misplaced or stolen. More than sixty federal, state, and local law enforcement agents descended on the university, and the media splashed his name across the news. Convicted of forty-four counts of research misconduct, Butler surrendered his medical license, lost his university appointment, and was sentenced to two years in federal prison.[22]

Post-9/11 biosecurity regulations increased the cost and the risk of research and development in sensitive areas of biology, thereby further reducing the likelihood of future industry and academic participation in vaccine development programs.[23] These regulations restrict the possession, use, and transfer of select agents and restrict the activities of foreign research scientists. They generate uncertainty and delays for research both within and outside of the United States. Rick Smith, director of Regulatory Affairs at Aventis Pasteur (now Sanofi-Aventis), observed that, whereas international collaboration was once relatively effortless, routine attempts to exchange seed strains between the United States, France, and Canada are now bogged down for weeks in paperwork and delays.[24] Michael Donneberg, associate chair of Microbiology and Immunology at the University of Maryland School of Medicine, argued that the additional bureaucratic and financial costs of conducting research on select pathogens "weighs into the question of whether to work with these agents."[25] Ronald Atlas, president of the American Society for Microbiology, echoed these sentiments, noting, "If I had select agents in my lab, I think I'd give serious consideration in the morning as to whether I really want to do this or not."[26] Given the professional risks and the uncertain cost and liability of working with select pathogens in a volatile and rapidly evolving regulatory environment, academic and industrial scientists are unlikely to opt for biodefense projects when viable research investment alternatives exist.

Even if these disincentives had not been present, it is not clear that large firms could have sustained a high level of assistance. Large manufacturers are not as free to respond to the call of patriotism as they were in the 1940s when they were stand-alone firms. All of the major vaccine firms have become part of larger publicly-owned companies that are highly sensitive to near-term earnings. After the vaccine industry consolidated, decision-making power was diffused not only among shareholders but also among nations. According to Walter Vandersmissen of SmithKline Beecham,

> All major vaccine companies have become global; operating from a European or American base has become irrelevant. That means that the industry, wherever it is based, is operating according to a number of principles, which have become virtually identical. Hence, economic, financial, and industrial feasibility are the main drivers that will determine what scientific or technical [projects] can be pursued . . . Vaccine manufacturers are no longer independent companies; they have become divisions within pharmaceutical companies where competition for resources is intense . . . To manage R&D, capital investment, [and] marketing expenditures to eventually come up with a new vaccine is now done with much less of a free hand than in the past.[27]

While the DOD struggled to get industry to accept vaccine development contracts, they also had difficulty getting service personnel to take the few vaccines they did have. In December of 1997, William Cohen, the secretary of defense, announced that the DOD would immunize all service personnel with the anthrax vaccine. Biodefense preventives and therapeutics were already under suspicion for having caused symptoms of the Gulf War syndrome so fears ran high among service personnel that the vaccine was unsafe, and several risked their military career by refusing the vaccine.

Congress fueled the fire. Referring to the DOD's anthrax immunization program, Congressman Christopher Shays (R-CT) ranted, "We're supposed to trust the military—and I wonder why, based on past experience, whether it was Agent Orange, whether it was people my office has had to help that have been exposed to radiation—we're supposed to trust the military to do the right thing . . . they have no basis in which to say, trust us."[28] For good measure, he added, "I'm not comfortable with generals practicing medicine, and I'm not comfortable with doctors planning wars, and, frankly, I'm not comfortable with doctors planning war doing medicine."[29]

Congress didn't place much faith in the pharmaceutical industry either. When Tommy Thompson asked Bayer to reduce the price of Cipro, Bayer's resistance spawned a vitriolic reaction. Senator Charles Schumer (D-NY) led the charge, stating, "we should not put our best response to anthrax in the hands of just one manufacturer."[30] Schumer identified five generic manufacturers who were willing and able to produce Cipro for the government and urged the FDA to approve them for this purpose. Just in case Bayer missed the message, Congressman Bernard Sanders (I-VT) explained: "When you have a national crisis, you do not have to give enormously profitable pharmaceutical companies the price they want. That is why we're here, to protect the American people, and if they want profits rather than serving the people, I think the law is very clear, that we have a right to go outside of that company [to break their patent]."[31]

~ AFTER SHOWCASING THE NEED for biodefense vaccines in the 2002 State of the Union Address, the Bush administration tried to resurrect biodefense efforts with a number of civilian push and pull programs, allocating $1.7 billion (annually) to the NIH for bioterrorism research and another $5.6 billion (over ten years) to encourage industry to develop biomedical countermeasures through Project BioShield.

Under these programs, the NIH overtook the DOD as the lead agency for directing civilian biodefense research and development (Table 6.1). Whereas the NIH had planned to spend $93 million on research related to bioterrorism prior to September 11, 2001, $1.7 billion was set aside for this purpose in FY 2003, nearly a 2,000 percent increase. According to Anthony Fauci, director of NIAID, this was "the biggest single-year request for any discipline or institute in the history of the NIH."[32] Fauci indicated he would set aside $441 million for basic research, $592 million for drug and vaccine discovery and development, and another $522 million to build new laboratories that could accommodate dangerous microbes. The rest of the funds, he explained, would be used to sponsor drug trials.[33]

This huge influx of funds is unlikely to have a significant effect on biodefense vaccine innovation rates, regardless of how Fauci chooses to distribute them. The NIH excels at generating new knowledge at the early stages of research, but it is fundamentally ill-suited for targeted research and development initiatives. NIH intramural and

Table 6.1 Medical Biodefense Funding in Millions

	FY00	FY01	FY02	FY03	FY04
HHS:NIH	43	49.7	275	1,700	1,600
DOD:[a]					
CBDP	363.5	377.6	569.5	598.1	559.6
DARPA	72.6	87.7	97.9	78.6	137.3
Total:	479.1	515	942.4	2,376.7	2,296.9
%Total NIH:	9%	10%	29%	72%	70%

a. Figures include the medical biodefense R&D component of the Chemical and Biological Defense Program and the Defense Advanced Research Projects Agency. The 2004 figures represent the president's FY2004 budget request.

Sources: AAAS R&D Analysis FY04, www.aaas.org/spp/rd; Institute of Medicine, *Giving Full Measure to Countermeasures: Addressing Problems in the DOD Program to Develop Medical Countermeasures* (Washington, DC: National Academies Press, 2004), 32–33.

extramural grant programs are highly specialized, and promotion is based on publications, not products. Consequently, NIH scientists are often less willing and able to bridge the basic-applied gap than their counterparts in military research institutions had been in previous decades. Unless the current shift of biodefense research and development resources from the military research labs to the NIH is met with some parallel shift in the governance structure for targeted vaccine development, numerous publications but few actual vaccines, will come out of NIH-sponsored biodefense programs in the future.[34]

In addition to funds to push new biodefense vaccines through the research phase, the Bush administration proposed $5.6 billion under Project BioShield to pull these vaccines through commercial development. Project BioShield was intended to (1) guarantee funding for a ten-year period to procure seven medical countermeasures (MCMs) to fight five pathogens; (2) authorize the FDA to make countermeasures under development available to the public in an emergency; and (3) bolster NIH resources to hire personnel, expedite peer review, and procure laboratory materials on short notice.[35]

Insulating the purchase fund from the annual appropriations process was heralded as a significant inducement to vaccine makers.[36] Five billion six hundred million dollars, however, is woefully insufficient to pull seven MCMs onto the market, let alone create a viable commercial biodefense industry.[37] According to the Congressional Budget Office, it will cost approximately $8.1 billion to develop seven MCMs over the next ten years, a 45 percent increase over the original White House estimate.[38] Subsequent estimates have been far higher, estimating $14 billion through FY 2015 to pull eight MCMs through with a 90 percent chance of success.[39]

The high failure rate of vaccine and drug candidates is a commonly cited cost driver. Rising estimates account for the need to fund multiple companies simultaneously to assure one successful outcome. Failure rates are not an independent variable, however. Low contract prices afforded under BioShield invite a higher failure rate because these contracts only appeal to small, inexperienced biotechnology companies in search of funding to stay afloat. BioShield margins were not to exceed 10 percent for individual contractors.

While these margins are standard in the defense industry, large experienced biopharmaceutical firms command margins closer to 28 percent to 31 percent in the commercial sector.[40] Since MCMs will be stockpiled, overall sales promise to be low as well. As a consequence, BioShield failed to attract the firms that were most likely to deliver licensed MCMs in a timely fashion.

Originally, legislators did not think that they would have to pay premium prices to engage competent manufacturers. The 2004 BioShield and 2006 BARDA legislation was designed to lure these companies with market guarantees, tax incentives, liability protections, accelerated regulatory review, and subsidized early-stage R&D. Because the government is in a position to reduce the cost and risk of vaccine development in this way, officials assumed they could negotiate lower contract prices. Experience has shown, however, that these inducements cannot compete with other opportunities in the market.

The opportunity costs for accepting government contracts are much higher today than they were in the 1940s and 1950s, when there were dozens of stand-alone vaccine companies. Now most developers are part of pharmaceutical conglomerates like Merck and Sanofi-Aventis that have a wider array of attractive product development and market opportunities. To meet manufacturing requirements for biodefense vaccine contracts, they would have to displace vaccine lines with viable commercial markets or invest in new facilities for biodefense vaccines. To date, they have been unwilling to do either.

As venture capital funding has become scarcer in recent years, a growing number of cash-starved biotechnology companies have accepted government biodefense contracts. In April 2002, the DOD planned a joint conference with the Biotechnology Industry Organization (BIO) to coordinate military needs with industry capabilities for biodefense. To their surprise, 360 biotechnology industry executives, twice the expected number, attended the briefing.[41] As Una Ryan, CEO of Avant, informed Congress at a hearing on biodefense vaccines, "We are not the pharmaceutical industry. We're small, we're nimble, we're unencumbered by profits, and we are extremely motivated to help the government."[42]

Many felt these smaller companies were the answer to the government's quest for affordable procurement options, but this avenue contained hidden costs and inefficiencies. Unlike large pharmaceutical companies, smaller biotechnology companies rarely have the expertise or infrastructure to perform late-stage vaccine development or manufacturing so they end up contracting out for these capabilities to fill government orders.[43] As chapter 5 illustrated, however, outsourcing can disrupt integrated research and development practices and the working relationships that support them.

The few integrated biotechnology companies that do exist are still relatively small and cannot handle large orders on their own. Acambis, for example, a biotechnology firm based in Cambridge, England, and Cambridge, Massachusetts, began producing limited quantities of the first cell-culture version of the smallpox vaccine for the CDC prior to 2001.[44] The CDC increased its order in the wake of the anthrax attacks, which forced Acambis to forge additional partnerships to fill the order. Acambis formed an alliance with the Baxter Healthcare Corporation to manufacture the vaccine in bulk in Austria, Chesapeake Biological Laboratories to package the vaccine in Baltimore, Maryland, and with the BioReliance Corporation to test the vaccine in Rockville, Maryland.

Similarly, BioPort (now Emergent BioSolutions), a small vaccine manufacturer in Michigan, won a contract to manufacture the DOD's anthrax vaccine even though the facility had failed FDA inspections in 1996 and failed multiple potency and sterility tests since 1998.[45] As BioPort struggled to meet the requirements for current Good Manufacturing Practices (cGMP), the DOD was forced to renegotiate costlier contracts and scale back its Anthrax Vaccine Immunization Program. It was not until BioPort contracted out its filling operations to another plant in Spokane, Washington, that the company was finally able to obtain FDA approval (a manufacturing supplement) in January 2002 at additional cost to the DOD.[46]

Cost overruns were not entirely due to incompetence. While BioPort/Emergent BioSolutions was inexperienced in vaccine development, it became skilled at using political contributions and lobbyists to negotiate outsized contracts. Emergent SEC 10-K filings revealed that their political handiwork produced markups of 300 percent for DOD BioThrax sales.[47] Company executives argue that the risks of vaccine development justify the markup. While Emergent did invest in making BioThrax cGMP compliant, they failed to acknowledge that most of the work on this vaccine had already been accomplished with taxpayer support over the last thirty years.[48]

BioThrax is not substantially different from the vaccine produced by Michigan Public Health labs in 1970. The current vaccine is still impractical for emergency use (requiring five shots over eighteen months to induce immunity), has adverse effects, suffers from lot-to-lot variation, and has a short shelf life. While the DOD was unwilling to pay a premium to engage a competent manufacturer up front, in the end it paid a premium for a product that was both substandard and late.

HHS struggled with biodefense contracts as well. The first large BioShield contract was a spectacular failure. In 2004, HHS awarded VaxGen (a small biotech company in California that had never successfully produced a licensed vaccine) a whopping $877 million to deliver a next-generation recombinant protective antigen (rPA) anthrax vaccine. HHS and Vaxgen contend that this contract, which required VaxGen to bring a Phase II candidate through licensure within two years, was overly ambitious and reflected wishful thinking on both sides.[49] When VaxGen ran into technical difficulties (for which it had no in-house expertise), it had to outsource the work. After a two-year extension, VaxGen was still struggling to resolve stability problems with the vaccine. Under pressure from lobbyists working for VaxGen's rival—Emergent BioSolutions—HHS canceled the contract, setting back the target date for a next-generation anthrax vaccine indefinitely. HHS underestimated the complexity of vaccine development and overestimated the manufacturing ability of small companies, much as the DOD did with BioPort.

HHS's protracted struggle to develop their highest priority countermeasure—a next-generation anthrax vaccine—illustrates how vaccines fall victim to the "valley of death." This phenomenon refers to the growing number of technologically feasible vaccines that fail for financial and/or organizational reasons. The majority of the cost and risk of vaccine development occurs in the late development stages with approximately 60 percent to 75 percent of the overall costs incurred during clinical trials and the start-up of the manufacturing phase.[50] Furthermore, only 20 percent of all drugs entering Phase 1 clinical trials are ever approved for commercial distribution.[51] While these numbers reflect some scientific obstacles, many are technical, organizational, and financial in nature. Small biotechnology companies rarely have the financial resources, facilities, or expertise to tackle these challenges on their own.[52]

Recognizing these advanced development challenges, Congress passed the Pandemic and All-hazards Preparedness Act (PAHPA, PL 109-417) in 2006. This legislation created the Biomedical Advanced Research and Development Authority (BARDA) to address many of the governance issues encountered under BioShield. In particular, BARDA was created to provide central oversight and financial flexibility for mid- and late-stage MCM development activities, allowing for milestone grants (up to 50 percent of the final contract) to be awarded when the company achieves prespecified milestones in the development process. This legislation also created the Public Health Emergency Medical Countermeasures Enterprise (PHEMCE) Governance Board to serve as the interagency body for coordinating and communicating MCM priorities and the National Biodefense Science Board (NBSB) to provide outside expert advice.

Milestone grants offer an immediate solution to the challenges facing small biotechnology companies trying to fill government contracts, but they do not solve the larger problem of engaging large experienced manufacturers that have a higher chance of success.[53] Under BARDA, HHS reissued a request for proposals (RFP) for another company to develop the rPA vaccine, only to discover that none of the applicants had the expertise to bring a vaccine—even one in the final stages of development—through licensure. In December 2009, BARDA canceled the RFP and issued a Broad Agency Announcement for smaller projects related to rPA vaccine development. In support of BARDA's shift toward milestone grants, Congress removed $884 million from BioShield's Special Reserve Fund in 2009, placing $304 million in NIAID research programs and $580 million with BARDA to manage late-stage development programs.[54]

These transfers undercut the BioShield's market guarantee, which was designed to help industry to manage its risk. With these moves, BARDA has retreated from the original concept of BioShield as a market-pull mechanism that encourages industry to assume the cost and risk of development in exchange for advance purchase guarantees. The entrepreneurial model envisioned in the 2004 BioShield legislation was a hands-off approach that allowed industry to pursue any development strategy of its choosing. Financial reward was contingent on a single endpoint—FDA licensure—regardless of how a company achieved it. Milestone grants depart from this model and require BARDA officials to micromanage the development process.

For the entrepreneurial model to work as intended, BARDA must offer larger procurement grants or expand the market. The first option is unlikely to work because awarding grants with a 31 percent margin to highly profitable pharmaceutical companies would be wildly unpopular, if not politically impossible, in Congress. BARDA is exploring the second option by encouraging members of the G-7 to purchase MCMs to build strategic stockpiles of their own. This avenue seems unlikely to work as well, given that there is little international agreement on the need for emergency stockpiles or on what they should contain.[55]

BioShield suffers from an additional problem that is more fundamental than its failure to attract competent manufacturers. BioShield is predicated on a fixed-defense stockpiling strategy (Chapter 1). Given the time, cost, and risk of vaccine development, it does not make sense to stockpile a long list of vaccines that can be quickly overcome by human engineering or natural evolution. Our powers of prediction are weak and the sheer number of possible threats will quickly outstrip our drug development resources.

The PHEMCE implementation plan under PAPHA recognizes this strategic limitation and aspires to a more flexible defense strategy, calling for "broad spectrum solutions" such as multiuse therapeutics, technologies, and processes that reduce the cost and time of development.[56] In support of this plan, in June of 2009, BARDA, solicited proposals to accelerate drug development tasks ranging from rapid diagnostics to new methods of bioprocess development and manufacturing, vaccine stabilization, and delivery.

While these are the types of investments that will eventually allow the United States to react to unforeseen threats more quickly and effectively, it is not clear that BARDA will be able to pursue this strategy over the long term. The PHEMCE implementation plan also pursues fixed defenses, calling for several pathogen-specific MCMs, and BARDA operates under an exceedingly constrained budget, which limits its ability to pursue both strategies simultaneously. In its first two years of operation, BARDA received under a third of the $1.7 billion authorized for fiscal years 2006–2008.

BARDA, like BioShield, has been underfunded, in part, because there has been relatively weak institutional and professional support for biodefense initiatives. Senator Bob Graham and Senator Jim Talent, chairman and vice chairman of the bipartisan Commission on the Prevention of Weapons of Mass Destruction Proliferation and Terrorism, point out that "There is no senior-level advocate for bio-preparedness in the Administration. Currently, there is a patchwork quilt of offices and agencies with more than two dozen Presidential-appointed, Senate-confirmed individuals with some oversight responsibility, but no single person in charge."[57] Without a high-level advocate, they argue, "no one focused on this priority during the debate on the stimulus legislation in February. As a result, the government missed an opportunity to adequately fund several biopreparedness initiatives including the Biomedical Advanced Research Development Authority or BARDA—one of the most important biodefense organizations in the United States."[58]

While BARDA has been underfunded, some would argue that NIH biodefense programs have been overfunded. In 2005, over 750 microbiologists signed an open letter to the NIH, arguing that the glut of new biodefense funding was corrupting research in the life sciences and ultimately threatening the health of the very public that biodefense programs were designed to protect.[59]

Referring to White House budget requests, President Bush explained, "It's money that we've got to spend. It's money that will enable me to say we're doing everything we can do to protect America."[60] Too often, however, the discussion of how to improve biodefense programs has centered on whether these programs are receiving too little money or too much. Instead, planners should devote more attention to the organization of the research and development process itself.

Vannevar Bush, seasoned veteran of federal research and development mobilization efforts, cautioned against any approach that involves too much money and too little coordination. Concerned by the expansion of federal research and development budgets during the Cold War, Bush warned, "If the country pours enough money into research, it will inevitably support the trivial and the mediocre. The supply of scientific manpower is not unlimited."[61] The key to success, he argued, lay not in the sum of money but in "the form of the organization" and the ability of "military officers, scientists, and engineers [to work] together effectively in partnership."[62]

Post-9/11 biodefense programs failed to heed this advice and demonstrated a poor understanding of the forces that drive vaccine innovation. The combination of push and pull programs currently in place will not bridge the valley of death because they reinforce the balkanized structure for research that grew in the 1980s and 1990s. Thus far, the government's preference for procuring goods through arm's-length transactions in the marketplace has only exacerbated efforts to achieve timely innovation.

In the following chapter, I outline a research and development program that will allow the United States to move toward a reactive model of biodefense in which MCMs can be developed in months rather than years. Far from being a whimsical aspiration, this is a strategic goal with a solid historical precedent. Many vaccines were produced in similar time frames during World War II without the benefit of twenty-first century process improvements. To accomplish this goal, however, the United States will have to devise an emergency MCM program that can withstand the vicissitudes of congressional support, administrative transitions, industrial support, and threat salience. The following chapter extracts lessons from the history of U.S. efforts to develop vaccines for national emergencies to inform these efforts going forward.

7 The Search for Sustainable Solutions

Biological weapons "received serious consideration on the part of all the combatants," cautioned Leroy Fothergill, technical director of the Chemical Warfare Service during World War II.[1] Fothergill predicted that enemy nations facing heavy industry and armament restrictions would "explore more subtle methods of warfare" by "prostitut[ing] facilities in the medical and biological fields for the purposes of developing biological warfare."[2] Despite these warnings, the national security community largely turned its attention away from biological threats after the 1960s.

By the turn of the twenty-first century, however, biological threats had become impossible to ignore. Fothergill's prediction proved correct, and several countries, the Soviet Union in particular, invested in biological weapons under the guise of biomedical research. The anthrax letter attacks and the subsequent emergence of SARS and novel flu strains also served as disturbing reminders that disease is an enduring, if not a growing, threat to national security and public health.

While pathogenic security threats have survived the test of time, our ability to respond to them has not. Our protracted effort to develop a next-generation anthrax vaccine is symptomatic of a growing national struggle to generate timely innovation. This example is particularly humbling since it is on the easy end of the spectrum of medical countermeasure (MCM) challenges; the rPA vaccine is a technologically feasible candidate that industry is able, if unwilling, to manufacture. If the Department of Defense (DOD) had persuaded Merck to accept the anthrax vaccine contract in 2001, for example, the United States would have licensed a new vaccine years ago.[3]

Unfortunately, few biodefense challenges are this straightforward. The most devastating pathogenic threats will be unforeseen and unfamiliar. Given the wide range of threats and our limited powers of prediction, extensive stockpiling is not merely wasteful but hubristic. The United States needs to build an emergency MCM program that will allow us to respond to new threats as they arise.

Midcentury vaccine development programs can inform these efforts, but they do not offer a perfect template for today. World War II vaccine development programs were cobbled together under emergency conditions. They pooled the talents of the best scientists and industrialists of their generation, focused their attention on targeted research and development objectives, and demanded products in a short time frame. Their efforts were heroic and their accomplishments were historic, but these programs were not designed to endure. Irvin Stewart, deputy director of the Office of Scientific Research and Development (OSRD) observed, "Once the pressure of war lifted, the key men upon whom its success depended responded to the more urgent calls of their regular activities and not all the king's horses nor all the king's men could hold the group together."[4]

Given the protracted and unpredictable nature of future disease threats and the long-range challenge of accelerating development times, a modern-day emergency MCM program must be sustainable if it is to succeed. Securing steady funds to support long-term research and development objectives will be a key challenge. While the market supports the development of some enabling technologies, it does not support the large-scale system building required to make them work. Efforts to integrate functions both before and after traditional drug development tasks, such as surveillance, detection, diagnosis, and distribution, will be more effective if they are supported and conducted outside of the private sector. This can be accomplished through a government laboratory or a quasi-government partnership such as a government-owned, company-operated (GOCO) or a public-private product development partnership (PDP).

One could argue that federal funding for a new vaccine initiative is not sustainable either given the current fiscal constraints of the U.S. government. It is important, however, to put this proposal into perspective. History demonstrates that purely private sector solutions are unreliable and some level of government investment is necessary to protect public health and national security objectives. Moreover, redirecting resources in this fashion could reduce the net cost of an emergency MCM program, since it would replace an expensive and lengthy stockpiling strategy that is likely to fail (Chapter 1).

Since this emergency MCM program will focus on process, rather than product, innovation, the economic argument for a market-based approach becomes weaker still. Industry makes money from new products, whereas process improvements may only improve productivity on the margin and often require regulatory validation.[5] Disincentives exist in academia too; scientists build their careers on publications, and they may have little to gain professionally from improving research tools. Redirecting federal resources to design an emergency countermeasure development system will provide the incentive structures and career paths that would build skills and spur investment in this critical, yet neglected, area of research.

It is not immediately obvious to historians, industry observers, or policy makers that the government should assume a more direct responsibility for vaccine development. Industry continued to respond to low-margin military vaccine requests and continued to generate high rates of innovation long after the United States dismantled wartime development programs. The productivity of the postwar era encouraged Louis Galambos and Jane Eliot Sewell to conclude that a mixed system of industry, government, and academic players, each of which responds to a different set of economic, political, and scientific incentives (effectively the status quo), generates the highest rate of innovation. They argue that this tripartite system, when left to its own devices, best serves the public's long-term interest in disease control.[6]

Studies based on publicly available vaccine license data support this conclusion and give the impression that the vaccine industry has effectively responded to economic and technological opportunities as they arose. However, as chapter 2 demonstrates, this interpretation is at odds with more historically accurate innovation patterns. Most notably, the current tripartite system has not responded to the public's long-term interest in global health (Table 1.1) and national security (Table 1.2).

This system generated high rates of innovation in the postwar period, but as chapters 4 and 5 reveal, success cannot be attributed to the inherent value of a tripartite system that ultimately allows corporations to decide which vaccines will be developed. This system was able to function ungoverned only insofar as the informal culture that supported military-industrial collaboration remained intact. During this period, collaborative networks fueled by patriotism, social obligation, familiarity, and trust offset the inherent lack of economic incentives for industry to participate in the development of socially valuable vaccines. What appeared, therefore, as a high-functioning system was in reality based on a uniquely productive partnership that began to unravel during the 1970s.

The 2001 anthrax letter attacks raised the visibility of biological threats and reestablished vaccines as an instrument of national security. In this post-9/11 atmosphere of urgency, the Bush administration requested large sums for vaccine research and development. Unlike World War II programs, however, the push and pull programs devised were inappropriate and insufficient to ensure timely innovation (Chapter 6).

Could industrial, academic, and government labs eventually self-organize as they did in the post World War II era? Political, institutional, legal, and economic transformations to the landscape for vaccine research and development (Chapter 5) make this outcome highly unlikely. Today, vaccine research scientists are more likely to hail from academia or the National Institutes of Health (NIH) than from a military lab, and they are less likely to possess the interdisciplinary skills required for targeted vaccine development. Even if they did have the proper training, today's science integrators face a more difficult task since vaccine development expertise has become more specialized and distributed among a balkanized set of institutions. Given the growing structural challenges to integrated research and development, emergency MCM programs are going to require more government support and direction than the current tripartite system allows.

Government ownership and/or control over product development is an old and unpopular idea. The repeated rejection of vaccine GOCO proposals in the United States (Chapter 5) reflects a pervading faith in the superiority of private-sector approaches and a desire to reduce the government's role as a producer. In 1998, Harvard economist Andrei Shleifer observed, "Today, the war and the depression are no longer as vivid, and the communist economies have collapsed. As importantly, the quality of contracting and regulation has improved, competition has become more effective, and the dangers of politicization of production have become self-evident, and the appreciation of the innovative potential of entrepreneurial firms is at a new high. For all these reasons, the benefits of reducing the role of government as a producer are becoming apparent and are beginning to be exploited."[7]

Shleifer perceived that the need for patriotic gestures had diminished and the cost of doing business was lower in the private sector than in the public sector. The prevailing view has been that—all things being equal—private firms offer higher-quality improvements (innovation) at a lower cost when owners are in a position to capture the returns on their investment.[8]

All things are not equal in vaccine development, however. Market competition is a demonstrated means of improving efficiency and innovation in many fields, but this model has not always worked as well for vaccines as it has for other technologies.[9] This is particularly true for vaccines of high social value (national security and/or public health), but low commercial appeal. As DOD and Health and Human Services (HHS) procurement efforts have demonstrated, the cost reduction, quality improvement, and innovation advantages inherent in market solutions do not apply to vaccines when the government fails to engage fully integrated and experienced manufacturers. High opportunity costs put integrated manufacturers out of a politically acceptable price range, and smaller companies carry hidden costs in the form of chronic contract renegotiations, supplemental subcontracting, longer development times, and higher failure rates.

Ideological opposition to GOCO proposals may have caused lawmakers to overlook their value in the case of biodefense vaccines. The question of whether to build a GOCO is akin to the routine "make versus buy" decisions that commercial firms face. These decisions weigh the relative transaction costs of developing a good in-house versus contracting for a good in the marketplace or developing it through mixed-mode relationships, such as joint ventures.[10] The level of risk inherent in particular projects and the level of trust between contracting parties influences transaction costs, which in turn determine the appropriate governance structure for product development.

Peter Ring and Andrew Van de Ven developed a model that incorporates risk and trust variables into transaction cost analyses to explain governance choices across the market-hierarchy spectrum.[11] While they developed this model to explain individual firm behavior, it sheds light on the changing nature of business/government arrangements to develop commercially unattractive vaccines. Risk, in this analogy, refers to the probability of bringing a countermeasure through licensure given a range of scientific, technological, regulatory, and market (in this case, government demand) uncertainties. Trust refers to "confidence in the other's goodwill."[12]

Ring and Van de Ven argue that when risk and trust are high, relational contracts (such as joint ventures) tend to work best. Relational contracts cover transactions that occur over an extended period of time. Under these circumstances, contracts can be awarded years in advance of development and licensure. The terms of exchange can be uncertain and open to ongoing negotiation, and disputes are resolved endogenously (i.e., often without appealing to market norms and lawsuits). "Trust" in this scenario "is the principal mode of social control among parties, using recurrent or relational contracting as a means of governance."[13] When trust is strong, contracting parties can afford informal and flexible contract arrangements, which are desirable given the risk and complexity of vaccine development.

Vaccine developers have to manage scientific and technological uncertainties, long development times, shifting regulatory requirements, and often uncertain government demand. Even so, high levels of trust characterized many midcentury military vaccine development ventures. Longstanding working relationships, personal friendships, and a revolving door between military and industrial labs bred trust and a sense of social obligation.[14] These collaborative networks employed freewheeling relational contracts with relatively few safeguards to enforce them. Collaboration in the 1950s and 1960s was highly productive and relatively unfettered by intellectual property and liability concerns.

When trust began to deteriorate during the 1970s, relational contracts also began to fail (Chapter 5). The cost and risk of vaccine development rose with the emergence of intellectual property (IP) thickets and larger clinical trials. Government and industry adopted more safeguards in their contracts and moved toward arm's-length market transactions to complete discrete pieces of the vaccine development puzzle. While these transactions looked efficient on paper, they did not work as well for vaccines as they did for military hardware acquisitions because it is more difficult to prespecify contracts for new vaccines with a high level of predictability.

Ring and Van de Ven observe that high-risk, low-trust transactions operate best under more hierarchical governance structures where "institutional role arrangements attenuate the dependency on personal trust characteristic of behavior in socially embedded relationships."[15] Under hierarchical arrangements, employment contracts and authority structures, rather than market norms and relational contracts, are used to resolve conflicts.

This is not, however, how the United States approached difficulties associated with vaccine development in the late twentieth century. Rather than trying to integrate research, development, and manufacturing activities vertically, industry and government migrated toward more horizontal arrangements. The 1980s and 1990s witnessed the emergence of small biotechnology firms as early innovators and the growth of lean organizations that contract for services in the marketplace. While this was a rational reaction to the explosion of specialized (and increasingly stovepiped) knowledge in the life sciences, vaccine development became increasingly disaggregated.

Since this high-risk, low-trust environment persists today, hierarchical governance structures may be a more prudent way to rejuvenate biomedical countermeasure development in the near term. According to Ring and Van de Ven, such structures eliminate the need for "complex and costly negotiations to anticipate properly the totality of investments and all possible outcome contingencies that may be entailed in undertaking a high-risk venture through a discrete contract-based market transaction."[16]

An emergency MCM program could provide the hierarchical structure required to navigate this new environment for MCM development. Despite the chronic unpopularity of GOCO proposals, building a biopharmaceutical facility in the national interest could address other important aspects of sustainability as well. Doing so could ultimately reduce the overall cost of development because the DOD or HHS would not have to pay a premium to engage qualified manufacturers. Most importantly, however, a government-sponsored facility would provide a stable environment to rebuild integrated research practices. Top-down control, integration across developmental phases and disciplines, and robust communities of practice are the three main ingredients of historically successful vaccine development programs. Building an emergency MCM development program offers a unique opportunity to reintroduce these research practices and to rebuild vaccine development capabilities.

While many midcentury research strategies have long since fallen out of fashion, James Conant's advice on research and development planning has as much merit today as it did during World War II. The former chairman of the OSRD's National Defense Research Committee concluded that the only way to advance pure science was to fund geniuses and allow their findings to trickle from the bottom up, whereas the only way to achieve results in applied science was to recruit the best talent in the field, give them a target, and manage research and development from the top down.[17]

A modern-day emergency MCM program will have to operate along the parallel tracks that Conant described. Some of the technologies and methods required to accelerate development times are in their infancy. HHS can designate funds toward early-stage research in this area within their preexisting bottom-up structures, which are well funded and high functioning. The NIH's new National Center for Advancing Translational Sciences (NCATS) promises to stimulate early-stage research in these critical areas. According to NIH Director Francis Collins, "NCATS's mission is to catalyze the generation of innovative methods and technologies that will enhance the development, testing, and implementation of diagnostics, therapeutics, and devices across a wide range of human disease and conditions."[18]

An emergency MCM program should employ technology watch teams to scan the horizon for relevant technologies and insights that emerge from these bottom-up processes and bring them in-house for further development and testing. The technology watch function is less passive than the name implies. These teams should also actively seek discoveries through new public-private consortia, such as the Biomarkers Consortium, and open-source innovation venues, such as InnoCentive. In many cases, it will be useful to bring the original discovery teams into emergency MCM program facilities to transfer their research and/or technology into this new context (Figure 7.1).

While discovery is well supported by the bottom-up structures and incentive systems that govern the NIH and academia, targeted programs to capture these discoveries and integrate them into an accelerated development system are lacking.[19] For these new discoveries and process innovations to bear fruit, they need a place to go to reach maturity. Rapid diagnostics, protein expression systems, and flexible trial designs (to name a few) must be incorporated into a broader system that will eventually be able to streamline the process from the identification of a novel pathogen to the development, testing, and distribution of new countermeasures. Even the newly proposed NCATS, which is intended to stimulate translational research, will explicitly "avoid a top-down management approach."[20] Industry is unlikely to provide the necessary leadership as well. Companies may adopt new processes that trickle up from these translational science initiatives in an incremental fashion, but they are unlikely to devise and implement a radical overhaul of drug development systems that would disrupt existing revenue streams.

To succeed, targeted development programs must support integrated research and development practices. Vaccine development remains an inherently interdisciplinary endeavor, requiring the in-

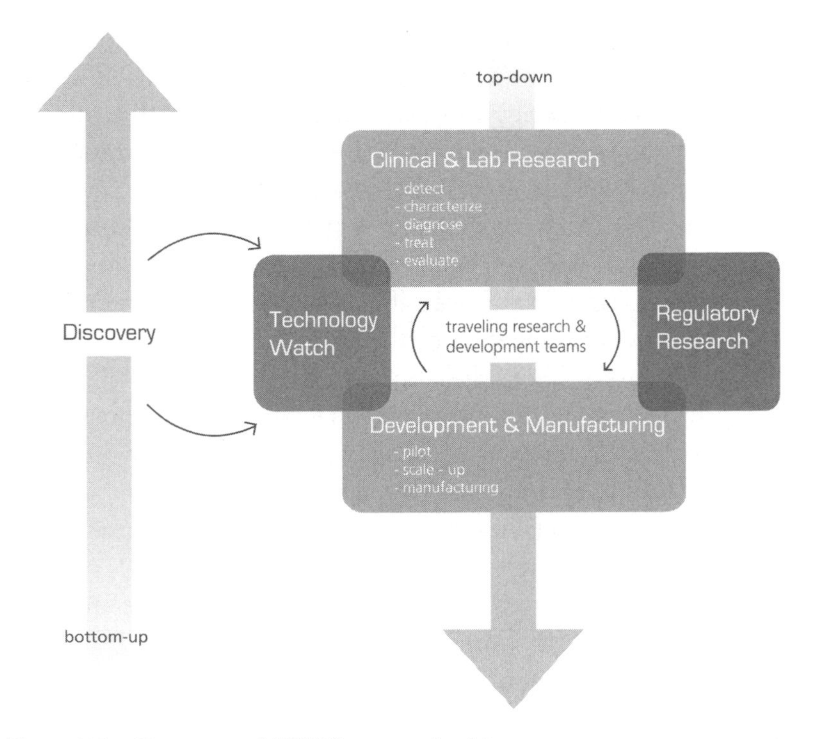

Figure 7.1: Emergency MCM Program Architecture

put of epidemiologists, laboratory scientists, regulatory scientists, and bioprocess engineers. World War II vaccine development programs and subsequent efforts at WRAIR, Merck, and National Drug succeeded when teams collaborated across disciplinary boundaries and development phases. This practice develops and preserves sticky knowledge and overcomes the type of technology transfer difficulties that have beleaguered MCM research and development initiatives in recent decades.

MCM development, and vaccine development in particular, is still tethered to a location and a community. Direct observation and on-site collaboration among developers, manufacturers, and user populations promotes effective technology transfer. This type of collaboration works best among individuals with mutual respect and repeated interactions. An emergency MCM facility would give sticky knowledge a place to live and grow and it would breed new communities of practice, much as WRAIR once did.

Hierarchical governance solutions are often interpreted as a decision to bring a function "in-house" or "under one roof." While the "under one roof" strategy can facilitate sticky knowledge transfer, World War II vaccine development programs demonstrated that it was not necessary to co-locate all research, development, and manufacturing activities, provided that a single development team or scientific director followed the vaccine across development phases. Organizing research in this way allowed academic and military researchers to work closely with epidemiologists, patient populations, and manufacturers. Similarly, top-down coordination in the industrial context allowed project director/scientists like Hilleman to coordinate research, clinical, and manufacturing teams with discovery teams and lead users in the field. Enabling project teams to travel from one development location to another will be essential to the success of this emergency MCM program.

Wartime project directors were also effective because they had the full support of the Surgeon General's Office (SGO) and the OSRD, which had both the perspective and the power to shift money, personnel, and/or equipment as necessary to facilitate high-priority projects. These offices, in turn, benefited from strong advisory bodies that coordinated the strategic objectives of the military with the technical and logistical capabilities of both academic scientists and industrial engineers. Close coordination among defense planners, research teams, and industrial planners ensured that operational needs, scientific hurdles, and manufacturing considerations were factored into research and development plans.

While World War II and WRAIR vaccine programs were highly targeted, they gave project directors a large degree of autonomy. Project directors were told what to develop but not how to develop it. The 2004 version of BioShield was intended to function the same way, but officials from the Biomedical Advanced Research and Development Authority (BARDA) have been unwilling to adopt the hands-off approach of their predecessors. The 2006 version of BioShield permits BARDA officials to micromanage MCM development through the use of milestone grants. By assigning discrete portions of the work in a piecemeal manner, BARDA officials take responsibility for product development out of the hands of those who are closest to the work and best able to chart a course for development. BARDA's growing reliance on milestone grants also requires them to use a tightly controlled set of carrot-and-stick incentives associated with "extrinsic motivation," which often impairs the ability of individuals and teams to arrive at creative solutions to ill-defined problems.[21]

Reinvesting scientists with the resources and authority to manage projects across the development spectrum is an important key to success. Not only would it support creative problem solving, but it would facilitate sticky knowledge transfer, improve situational awareness of product development needs, and permit timely decision-making. In this sense, integrated research is similar to the military concept of Command and Control.

The DOD defines Command and Control as "[t]he exercise of authority and direction by a properly designated commander over assigned and attached forces in the accomplishment of the mission."[22] The objective is not unilateral control but coordinated action that adapts to on the ground reality in a manner that is consistent with overriding strategic objectives. Effective command and control, therefore, depends heavily on the feedback, insight, and initiative of subordinate teams.

The command and control analogy captures the tension between the need to target research objectives and the importance of allowing investigators to use creative initiative and to implement tacit knowledge. It also emphasizes the importance of situational awareness in uncertain (research) environments to improve timely "go, no-go" decisions. Just as war can be viewed as a dynamic learning environment, so too can vaccine research. This is especially true if the research objective is to accelerate development times. Military commanders and research directors alike contend with on-the-ground information that is difficult to interpret out of context. They must optimize interoperability and situational awareness to ensure timely decision making. Effective command and control allows adaptive tempo changes that confer strategic advantages, whether the adversary is an insurgent or a rapidly evolving microbe.

This analogy also provides a framework for understanding how governance structures can shift over time to accommodate changes in the research environment. Just as military command and control systems must adjust top-down control to pursue greater and lesser degrees of certainty, so too must integrated research systems.[23] While the OSRD and the SGO imposed strong top-down direction during World War II, less direction was required during the postwar period when integrated research practices and collaborative networks were high functioning. As trust diminished and the risk and complexity of vaccine development increased, networks disintegrated and integrated research became more difficult to practice, an indication that vaccine development could benefit from greater top-down control once again. Over time, our approach to vaccinology may become less empirical and more rational, and other trust and risk variables may improve, which would allow us to migrate back towards more horizontal governance structures.

Emergency MCM project directors will function at times both as developers and as systems integrators—testing, validating, and linking prototype technologies and systems together. One can argue that these two functions should be kept separate. On the one hand, separation prevents the system integrators from becoming too invested in a particular technology and locked in to a suboptimal trajectory. On the other hand, the most objective party can also be the most uninformed. As the DynPort case demonstrates, this can spell disaster for vaccine development. Emergency MCM project directors may be more attracted to technologies and approaches that are similar to in-house projects, but this hands-on experience will equip them to make informed purchasing decisions about emerging component technologies in the commercial sector. Bias may also be less of a concern if this facility operates as a nonprofit and is therefore not compelled to protect a business base.[24]

It will be important for emergency MCM project directors to have the authority to manage their own projects and to maintain close ties to high-level planning groups and lead-user populations. In their study of the defense industry, Peter Dombrowski and Eugene Gholtz argue that systems integration is not simply a matter of ensuring interoperability, but of linking performance to strategic goals and making decisions about performance tradeoffs in the overall system architecture.[25] The ideal systems integrator, therefore, should (1) engage in development projects of its own so it can evaluate contractees on the basis of firsthand knowledge; (2) understand how military and civilian populations will use these technologies in practice; and (3) be a part of high-level strategic discussions about how MCMs will be used on the battlefield and in homeland defense scenarios.[26]

Successful midcentry vaccine development projects had one other important feature in common: reliable access to late-stage development and manufacturing facilities. Transformations to the landscape for vaccine development have constrained government access to these facilities and capabilities in industry, which is a leading cause for the valley of death phenomenon (Chapter 6).

In an effort to improve access to these facilities, a 2010 review of the Public Health Emergency Medical Countermeasures Enterprise (PHEMCE) implementation plan recommended building a government-sponsored manufacturing facility for MCMs. Without invoking the term "GOCO," HHS has requested proposals from potential industry and academic partners to build a center for innovation in advanced development and manufacturing (ADM).[27] The idea behind a public-private partnership of this nature is to draw on industry's manufacturing skills, the expertise of an academic research center, and federal funding to supplement development costs when the social returns from developing a particular vaccine exceed the financial returns.

Unfortunately, HHS's ADM request solicits skills that are not readily available. A dearth of late-stage development and manufacturing expertise has complicated federal efforts to launch expanded targeted vaccine development initiatives in the past. Anna Johnson-Winegar, former deputy assistant to the secretary of defense on Chemical and Biological Defense, noted that "the biotechnology and pharmaceutical industry as a whole is facing shortages of skilled personnel."[28] This problem is a direct result of industry contraction in the 1970s and 1980s.[29] Biomedical research and training has also become more specialized and stovepiped, causing fewer researchers to receive the type of hands-on, cross-disciplinary training required to develop new systems and products.

An emergency MCM program will have to rebuild a work force of "science integrators" through training and education, much like WRAIR once did.[30] This program will also create a viable career path for scientists and engineers pursuing interdisciplinary work and offer industry-competitive salaries.

While ADMs take a promising step toward integrating development activities such as pilot development, scale-up, and manufacturing, they do not go far enough. An emergency MCM program must integrate clinical and lab activities as well. One way to do this is to expand on the NIH's concept of Vaccine Treatment and Evaluation Units (VTEUs). Through VTEUs, the NIH partners with large academic health centers to enroll large numbers of volunteers into clinical trials. These trial centers played a key role in efforts to rapidly evaluate an H1N1 vaccine for the 2009 flu season.

Under their current configuration, VTEUs could play a critical role in the rapid evaluation of emergency MCMs. However, much can be gained from expanding the VTEU concept further to integrate clinical testing and evaluation with other epidemiological and laboratory activities such as disease tracking, diagnostic development, and target identification (Figure 7.1). WRAIR's vaccine development projects were most effective when Epidemiological Consultation (EPICON) teams were able to collaborate closely with laboratory researchers on site, for they enabled both parties to coordinate clinical, environmental, and laboratory requirements. Integrating clinical and lab functions at VTEU sites would facilitate efforts to detect and diagnose pathogen-borne illnesses, characterize unknown pathogens, identify novel biomarkers, develop and employ new diagnostics, devise faster clinical trial designs, and rapidly test MCM prototypes.

Last, but not least, it is essential to integrate regulatory research into the emergency MCM program. Unlike vaccines, which are typically given to healthy populations, emergency MCMs will be used in postattack scenarios where the risk of infection could outweigh risk of taking an investigational new drug or vaccine. Developers will need to work with regulatory scientists to establish rapid validation methods and with physicians to develop emergency-use protocols for this scenario. New diagnostic tools, biomarkers, adjuvants, manufacturing processes, and delivery systems will have to be developed in cooperation with regulators who understand the requirements of an emergency development system. To this end, Center for Biologics Evaluation and Research (CBER) teams need to establish research programs and maintain labs at emergency MCM program development sites to inform and evaluate development efforts earlier in the process than is typical for nonemergency drugs and vaccines.

Integrating research and development in this manner will remove organizational obstacles to vaccine development and introduce new efficiencies into the development process. When epidemiologists, clinicians, lab scientists, bioprocess engineers, and regulators are united under a single research objective (i.e., accelerating development times) and when they are intimately familiar with upstream and downstream requirements of their collaborators, they will begin to visualize and solve problems in new ways. For example, integration creates incentives to design detection systems that provide critical diagnostic information, diagnostics that facilitate the search for drug targets, and MCM candidates that are easily adapted to manufacturing platforms and validated for testing. Unlike the Joint Vaccine Acquisition Program (JVAP) or BARDA, this program will retain an institutional memory for approaches that do and do not work. Traveling teams and on-site learning afforded through this program will facilitate sticky-knowledge transfer.

A robust emergency MCM program will also build research communities that reunite lead users with product developers. Integrating clinical and lab activities in the context of academic medical facilities, like VTEUs, puts developers into direct contact with civilian lead-user populations such as health care workers. It will be important to maintain integrated clinical and lab facilities in military settings as well. Due to unique living and working conditions and the need for geographic flexibility, the military will always have protection needs that exceed those of the general population and a predisposition to invest in preventive measures. The military also continues to have unique infrastructure advantages, such as specialized facilities to study dangerous pathogens and their network of international labs, which offer tremendous advantages to researchers who need to work closely with populations most directly affected by disease.

Despite budget cuts to, and mismanagement of, the military laboratory system, WRAIR and USAMRIID researchers continue to make significant contributions to vaccine development. In the 1980s, WRAIR worked with SmithKline Beecham to license the first hepatitis A vaccine. In the 1980s and 1990s, DOD overseas labs organized clinical trials that led to the licensure of a next-generation Japanese encephalitis vaccine and that defined safety and efficacy for typhoid and cholera vaccines.[31] The underlying research and technology used for next-generation anthrax and smallpox vaccines (the rPA vaccine and the ACAM2000, respectively) also originated in military labs. The military is heavily invested in malaria and HIV vaccine research as well, and new developments are likely to emerge from their investments in these areas.

Finally, wartime programs succeeded because they motivated the best minds of their generation. A world war is a powerful recruiting tool. A peacetime program, by contrast, will have to brand itself as the place where the best minds go to tackle the most pressing global health problems as WRAIR once did and as places like the Gates Foundation do today. Restructuring the vaccine and drug development process to accelerate development times is an essential but immodest goal and it cannot be accomplished by bureaucratic fiat. It will be important to choose someone to lead this effort who, like Vannevar Bush or Maurice Hilleman, can attract talent, build support, form alliances, and effect change. This program must also provide viable career paths and confer prestige on the participating clinicians, scientists, and engineers.

Constructing systems that will allow us to respond to emerging biological threats more quickly may seem like a long shot—particularly when one considers that our efforts to develop a next-generation anthrax vaccine are over twenty years old—but this project is not without historical precedent. World War II vaccine development programs were remarkably productive and responsive to emergency needs and flu manufacturers already develop vaccines for new strains on a six-month time line every year. These response times were possible even before developers began to exploit currently available process improvement technologies (i.e., reverse genetics and cell culture techniques), much less invest in new ones. Reintroducing integrated research practices and shifting the strategic focus of biodefense programs will address problems that have beset vaccine development for the past several decades and restore our capacity to respond to a rapidly evolving array of threats to global health and security.

Appendix 1

Vaccine License Data, 1903–1999

Year	Manufacturer	Product Name	Approved	Revoked	Innovation type	Notes
1903	Bayer Corporation	Diphtheria Antitoxin	08/21/1903	11/30/1970	I	Bayer listings were originally licensed to Cutter Labs in Berkeley, CA. Bayer did not acquire Cutter until 1974.
1903	Bayer Corporation	Smallpox Vaccine	08/21/1903	06/11/1973	I	Parke-Davis, Frederick Stearns, National Vaccine Establishment, Fluid Vaccine, Mulford, and New York Bureau of Health also held smallpox licenses, but original approval dates are unavailable. Arthur Allen, *Vaccine* (New York: Norton, 2007), 96.
1903	Pocono Laboratories	Smallpox Vaccine	08/21/1903	07/01/1908	I	Jeff Widmer, *The Spirit of Swiftwater: 100 Years at the Pocono Labs* (Swiftwater, PA: Connaught Laboratories, 1997).
1903	Wyeth Laboratories	Smallpox Vaccine	08/12/1903		I	Alexander held this original license. Alexander was taken over by Wyeth in 1942. Office of Technology Assessment (OTA), *A Review of Selected Federal Vaccine and Immunization Policies: Based on Case Studies of Pneumococcal Vaccine* (Washington,

						DC: U.S. Government Printing Office 1979), Appendix 2: "Profile of vaccine establishments and products currently licensed in the U.S. (1979)."
1907	Parke-Davis	Diphtheria Antitoxin	11/15/1907	11/24/1969	I	
1907	Parke-Davis	Tetanus Antitoxin	11/15/1907	03/21/1973	I	
1908	Lederle Laboratories	Diphtheria Antitoxin	00/00/1908	01/19/1914	I	PHS, *Biological Products*, Publication 50, 1908.
1908	Lederle Laboratories	Tetanus Antitoxin	00/00/1908		I	PHS, *Biological Products*, Publication 50, 1908.
1908	Lederle Laboratories	Typhoid Vaccine	00/00/1908		C	PHS, *Biological Products*, Publication 50, 1908.
1911	Cutter Biological	Plague Vaccine	00/00/1911	05/14/1942	C	
1912	Slee Laboratories	Diphtheria Antitoxin	00/00/1912	12/12/1927	I	*Licensed Biologicals*, Slee Laboratories, 1912, AP Archives.
1912	Slee Laboratories	Smallpox Vaccine	07/12/1912		LT	Transferred from Pocono Labs. *Licensed Biologicals*, Slee Laboratories, 1912, AP Archives.

Vaccine License Data, 1903–1999 (continued)

Year	Manufacturer	Product Name	Approved	Revoked	Innovation type	Notes
1912	Slee Laboratories	Tetanus Antitoxin	00/00/1912		I	*Licensed Biologicals*, Slee Laboratories, 1912, AP Archives. Slee Laboratories also produced gas gangrene antitoxin for the military, 1913–1914, although there is no record of a license.
1914	Bayer Corporation	Pertussis Vaccine	08/03/1914	10/30/1970	C	
1914	Bayer Corporation	Tetanus Antitoxin	08/03/1914	10/30/1970	I	
1914	Lederle Laboratories	Diphtheria Antitoxin	01/19/1914	06/13/1977	LT	
1914	Lederle Laboratories	Pertussis Vaccine	01/19/1914	05/29/1980	C	
1914	Lederle Laboratories	Tetanus Antitoxin	01/19/1914	05/05/1976	I	
1914	Parke-Davis	Pertussis Vaccine	00/00/1914	04/16/1952	C	
1914	Wyeth Laboratories	Cholera Vaccine	00/00/1914		C	PHS, *Biological Products*, Publication 50, 1914.
1914	Wyeth Laboratories	Typhoid Vaccine	00/00/1914		I	For military use only; acetone-killed and dried.
1914	Wyeth Laboratories	Typhoid Vaccine	00/00/1914		I	
1915	Eli Lilly Laboratory	Pertussis Vaccine	03/31/1915	03/07/1978	I	
1915	Eli Lilly Laboratory	Rabies Vaccine	06/07/1915	08/11/1982	I	

1915	Parke-Davis	Tetanus Antitoxin	01/13/1915		I	OTA, *A Review of Selected Federal Vaccine and Immunization Policies*, (1979), Appendix 2: "Profile of vaccine establishments and products currently licensed in the U.S. (1979)."
1916	Bayer Corporation	Typhoid Vaccine	03/06/1916	10/30/1970	I	
1916	Parke-Davis	Typhoid Vaccine	03/24/1916	02/24/1969	I	
1917	Eli Lilly Laboratory	Cholera Vaccine	10/31/1917	06/07/1979	I	Not significantly different from the original Kolle vaccine.
1917	Parke-Davis	Pertussis Vaccine	00/24/1917		I	OTA, *A Review of Selected Federal Vaccine and Immunization Policies*, (1979), Appendix 2: "Profile of vaccine establishments and products currently licensed in the U.S. (1979)."
1917	MPHBL	Diphtheria Antitoxin	03/20/1917	10/26/1988	I	MPHBL (Massachusetts Public Health Biological Laboratories).
1917	MPHBL	Smallpox Vaccine	03/20/1917	12/22/1976	I	
1917	MPHBL	Typhoid Vaccine	03/20/1917	10/26/1988	I	
1918	Parke-Davis	Perfringens Antitoxin	09/23/1918	03/21/1973	C	

Vaccine License Data, 1903–1999 (continued)

Year	Manufacturer	Product Name	Approved	Revoked	Innovation type	Notes
1926	Merrell-National Laboratories	Pertussis Vaccine	10/16/1926	01/03/1978	I	The National Drug Company bought Slee Labs in 1926. National did not merge with Merrell-Richardson until 1961, but CBER lists National in its merged form since 1926.
1926	BLMDPH	Diphtheria Antitoxin	05/17/1926	05/11/1987	I	BLMDPH (Bureau of Laboratories, Michigan Department of Public Health).
1926	BLMDPH	Typhoid Vaccine	07/26/1926	06/25/1985	I	
1927	Merrell-National Laboratories	Diphtheria Antitoxin	12/12/1927	01/03/1978	LT	
1927	Merrell-National Laboratories	Tetanus Antitoxin	12/14/1927	01/03/1978	LT	
1927	Merrell-National Laboratories	Typhoid Vaccine	01/14/1927		I	The National Drug Company, *Biologics by National* (1956), AP Archives.
1927	Parke-Davis	Diphtheria Toxoid	08/17/1927	10/14/1981	P	Active diphtheria immunization.
1928	Bayer Corporation	Diphtheria Toxoid	02/01/1928	10/30/1970	I	
1929	Bayer Corporation	Perfringens Antitoxin	06/14/1929	06/14/1965	I	

1929	Merrell-National Laboratories	Diphtheria Toxoid	09/28/1929	01/03/1978	I	
1930	BLMDPH	Smallpox Vaccine	10/01/1930		I	
1932	Dow Chemical Company	Pertussis Vaccine	10/01/1932	06/07/1977	I	
1932	MPHBL	Diphtheria Toxoid	07/07/1932	05/19/1980	I	
1933	Lederle Laboratories	Staphylococcus Toxoid	04/03/1933	05/21/1980	C	
1933	Merck & Company	Tetanus Toxoid	12/11/1933	01/31/1986	P	Active tetanus immunization.
1933	Parke-Davis	Tetanus Toxoid	00/00/1933	09/25/1940	P	A. Beneson, "Immunization and Military Medicine," *Review of Infectious Diseases*, vol. 6, no. 1 (1984) 1–12.
1934	Dow Chemical Company	Diphtheria Toxoid	09/18/1934	06/07/1977	I	
1934	Merrell-National Laboratories	Tetanus Toxoid	05/25/1934	01/03/1978	I	
1934	Merrell-National Laboratories	Tetanus Toxoid Adsorbed	05/25/1934	01/03/1978	P	First licensed use of alum adjuvant, improved immune response. A. Beneson, "Immunization and Military Medicine," *Review of Infectious Diseases*, vol. 6, no. 1 (1984) 1–12; The National Drug Company, *Biologics by National.*

Vaccine License Data, 1903–1999 (continued)

Year	Manufacturer	Product Name	Approved	Revoked	Innovation type	Notes
1935	Eli Lilly Laboratory	Tetanus Toxoid	12/10/1935	06/07/1979	I	
1935	Lederle Laboratories	Tetanus Toxoid	06/15/1935	03/04/1994	I	
1935	BLMDPH	Pertussis Vaccine	11/22/1935	02/03/1977	I	
1936	Dow Chemical Company	Tetanus Toxoid	08/01/1936	06/07/1977	I	
1936	Parke-Davis	Staphylococcus Toxoid	06/06/1936	03/21/1973	I	
1936	Texas Department of Health Resources	Diphtheria Toxoid	01/06/1936	02/06/1979	I	
1937	Lederle Laboratories	Smallpox Vaccine	03/01/1937	05/24/1978	I	
1937	Merrell-National Laboratories	Smallpox Vaccine	03/05/1937	01/03/1978	I	
1937	BLMDPH	Smallpox Vaccine	03/03/1937	06/25/1985	I	
1937	Parke-Davis	Staphylococcus Antitoxin	11/27/1937	08/08/1944	I	
1940	Bayer Corporation	Tetanus Toxoid	09/25/1940	11/01/1979	I	
1940	Parke-Davis	Tetanus Toxoid	05/04/1940	10/14/1981	I	
1941	Sharp & Dhome	Typhus Vaccine	12/24/1941		P	OTA, *A Review of Selected Federal Vaccine and Immunization Policies,*

						(1979), Appendix 2: "Profile of vaccine establishments and products currently licensed in the U.S. (1979)."
1941	Lederle Laboratories	Botulism Antitoxin	10/19/1941	03/12/1981	C	
1941	Lederle Laboratories	Cholera Vaccine	12/26/1941	10/23/1996	I	
1942	Bayer Corporation	Plague Vaccine	05/14/1942	05/24/1995	C	Whole cell, formalin inactivated. Earlier vaccine of unproven efficacy. According to one account, the U.S. approved an "anti-plague" vaccine from Swiss Serum and Vaccine Institute Berne in 1908. R. Rader, *Biopharma: Biopharmaceutical Products in the U.S. Market* (Rockville, MD: Biotechnology Information Institute, 2001) 233.
1942	Eli Lilly Laboratory	Typhus Vaccine	00/00/1942	06/07/1979	I	W. J. Draper, acting surgeon general, to Senator David Walsh, February 4, 1942, NA: RG 443, E. 1, Decimal file #0470–132.
1942	Lederle Laboratories	Typhus Vaccine	00/00/1942	00/00/1967	I	W. J. Draper, acting surgeon general, to Senator David Walsh, February 4, 1942, NA: RG 443, E. 1, Decimal file #0470–132.

Vaccine License Data, 1903–1999 (continued)

Year	Manufacturer	Product Name	Approved	Revoked	Innovation type	Notes
1942	Sharp & Dohme	Typhus Vaccine	00/00/1942		I	W. J. Draper, acting surgeon general, to Senator David Walsh, February 4, 1942, NA: RG 443, E. 1, Decimal file #0470–132.
1942	Lederle Laboratories	Rocky Mountain Spotted Fever Vaccine	04/13/1942	06/11/1979	C	The original license was likely issued to Rocky Mountain Laboratories (RML). RML also produced a typhus and yellow fever vaccine for the military during this period. RML licenses were transferred to Lederle.
1942	Merrell-National Laboratories	Cholera Vaccine	02/27/1942	03/25/1976	I	
1942	Parke-Davis	Rabies Vaccine	08/05/1942	03/21/1973	I	
1942	Parke-Davis	Typhus Vaccine	03/25/1942	08/05/1947	I	
1944	Wyeth Laboratories	Pertussis Vaccine	05/19/1944		I	OTA, *A Review of Selected Federal Vaccine and Immunization Policies*, (1979), Appendix 2: "Profile of vaccine establishments and products currently licensed in the U.S. (1979)."
1944	Wyeth Laboratories	Diphtheria Toxoid	05/19/1944	09/11/1970	I	

1944	Wyeth Laboratories	Smallpox Vaccine	05/19/1944	00/00/2008	P	Lyophilized calf lymph vaccine (Dryvax). Wyeth ceased commercial manufacture in 1982 but maintained the license until 2008.
1944	Wyeth Laboratories	Typhoid Vaccine	05/19/1944		I	OTA, *A Review of Selected Federal Vaccine and Immunization Policies*, (1979), Appendix 2: "Profile of vaccine establishments and products currently licensed in the U.S. (1979)."
1944	Wyeth Laboratories	Tetanus Toxoid	05/19/1944		I	
1945	Eli Lilly Laboratory	Influenza Virus Vaccine	11/09/1945	04/11/1977	C	
1945	Lederle Laboratories	Influenza Virus Vaccine	12/07/1945		C	Flu-immune; split viron.
1945	Sharp & Dohme	Influenza Virus Vaccine	11/30/1945	03/15/1995	C	
1945	Parke-Davis	Influenza Virus Vaccine	11/26/1945	04/20/1998	C	Fluogen; split viron.
1946	Parke-Davis	Diphtheria and Tetanus Toxoid, Pertussis Vaccine Adsorbed	01/30/1946		A	OTA, *A Review of Selected Federal Vaccine and Immunization Policies*, (1979), Appendix 2: "Profile of vaccine establishments and products currently licensed in the U.S. (1979)."

Vaccine License Data, 1903–1999 (continued)

Year	Manufacturer	Product Name	Approved	Revoked	Innovation type	Notes
1947	Merrell-National Laboratories	Influenza Virus Vaccine	09/16/1947	01/03/1978	I	Purified with Sharples centrifuge, grown in chick embryo. National also manufactured an unlicensed Japanese encephalitis vaccine under government contract.
1948	Bayer Corporation	Pertussis Vaccine Adsorbed	09/03/1948	10/30/1970	P	Improved immunogenicity.
1948	Lederle Laboratories	Diphtheria and Tetanus Toxoid, Pertussis Vaccine Adsorbed	03/15/1948		I	OTA, *A Review of Selected Federal Vaccine and Immunization Policies,* (1979), Appendix 2: "Profile of vaccine establishments and products currently licensed in the U.S. (1979)."
1948	BLMDPH	Diphtheria and Tetanus Toxoid, Pertussis Vaccine Adsorbed	05/13/1948		I	OTA, *A Review of Selected Federal Vaccine and Immunization Policies,* (1979), Appendix 2: "Profile of vaccine establishments and products currently licensed in the U.S. (1979)."
1948	Sharp & Dohme	Rocky Mountain Spotted Fever	00/00/1948		I	Leonard Scheele, surgeon general, to Dr. Underwood, secretary, State Board of Health, Mississippi, September 27, 1948, NA: RG 443,

E. 1, Decimal file #0470.

1948	Squibb & Sons	Pneumococcal Vaccine, Polyvalent	00/00/1948	00/00/1954	R	First active immunization against pneumococcal pneumonia (C) and first polysaccharide vaccine (P). M. Heidelberger, "A 'Pure' Organic Chemist's Downward Path," *Annual Review of Biochemistry* 48 (1979): 1–21.
1949	Bayer Corporation	Diphtheria and Tetanus Toxoids Adsorbed	05/04/1949	10/30/1970	A	
1949	Bayer Corporation	Diphtheria and Tetanus Toxoids and Pertussis Vaccine	05/04/1949	10/30/1970	A	
1949	Bayer Corporation	Diphtheria and	05/04/1949	10/30/1970	I	
1949	Bayer Corporation	Diphtheria Toxoid Adsorbed Tetanus Toxoids and Pertussis Vaccine Adsorbed	05/04/1949	10/30/1970	P	Improved immune response. First recorded use of alum in diphtheria toxoid according to CBER, although some army records indicate earlier use of adsorbed diphtheria toxoids.
1949	Bayer Corporation	Diphtheria Toxoid and Pertussis Vaccine Adsorbed	05/04/1949	10/30/1970	I	

Vaccine License Data, 1903–1999 (continued)

Year	Manufacturer	Product Name	Approved	Revoked	Innovation type	Notes
1949	Bayer Corporation	Gas Gangrene Polyvalent Antitoxin	05/04/1949	06/14/1965	A	A. Beneson, "Immunization and Military Medicine," *Review of Infectious Diseases*, vol. 6, no. 1 (1984) 1–12.
1949	Bayer Corporation	Tetanus and Gas Gangrene Polyvalent Antitoxin	05/04/1949	06/14/1965	A	
1949	Bayer Corporation	Tetanus Toxoid Adsorbed	05/04/1949	10/30/1970	I	Some records indicate that the first tetantus toxoid absorbed was licensed in 1937.
1949	Wyeth Laboratories	Diphtheria and Tetanus Toxoids Adsorbed	07/26/1949	07/26/1970	A	OTA, *A Review of Selected Federal Vaccine and Immunization Policies*, (1979), Appendix 2: "Profile of vaccine establishments and products currently licensed in the U.S. (1979)."
1949	Eli Lilly Laboratories	Diphtheria and Tetanus Toxoids	07/26/1949		I	OTA, *A Review of Selected Federal Vaccine and Immunization Policies*, (1979), Appendix 2: "Profile of vaccine establishments and products currently licensed in the U.S. (1979)."

1949	Eli Lilly Laboratories	Diphtheria and Tetanus Toxoids Adsorbed	07/26/1949		I	OTA, *A Review of Selected Federal Vaccine and Immunization Policies*, (1979), Appendix 2: "Profile of vaccine establishments and products currently licensed in the U.S. (1979)."
1949	Sharp & Dohme	Diphtheria Toxoid and Pertussis Vaccine Adsorbed	03/31/1949		I	OTA, *A Review of Selected Federal Vaccine and Immunization Policies*, (1979), Appendix 2: "Profile of vaccine establishments and products currently licensed in the U.S. (1979)."
1949	Lederle Laboratories	Gas Gangrene Polyvalent Antitoxin	05/04/1949	03/12/1981	A	OTA, *A Review of Selected Federal Vaccine and Immunization Policies*, (1979), Appendix 2: "Profile of vaccine establishments and products currently licensed in the U.S. (1979)."
1949	Lederle Laboratories	Tetanus and Gas Gangrene Polyvalent Antitoxin	05/04/1949	05/05/1976	A	
1949	MPHBL	Tetanus Toxoid	05/16/1949	10/11/1989	I	
1949	Wyeth Laboratories	Diphtheria and Tetanus Toxoids Adsorbed	07/26/1940		I	OTA, *A Review of Selected Federal Vaccine and Immunization Policies*, (1979), Appendix 2: "Profile of vaccine establishments and products currently licensed in the U.S. (1979)."

Vaccine License Data, 1903–1999 (continued)

Year	Manufacturer	Product Name	Approved	Revoked	Innovation type	Notes
1949	Merrell-National Laboratories	Diphtheria and Tetanus Toxoids and Pertussis Vaccine Adsorbed	05/16/1949	01/03/1978	P	In 1944, Dr. Bolyn at National Drug used a smaller fraction of alum as precipitant for toxoid; other preparations used larger amounts (1/2%), which produced greater side effects. Jeff Widmer, *The Spirit of Swiftwater: 100 Years at the Pocono Labs* (Swiftwater, PA: Connaught Laboratories, 1997).
1949	Parke-Davis	Tetanus and Gas Gangrene Polyvalent Antitoxin	04/08/1949	03/21/1973	A	
1949	Parke-Davis	Diphtheria and Tetanus Toxoids Adsorbed	04/08/1949		I	OTA, *A Review of Selected Federal Vaccine and Immunization Policies*, (1979), Appendix 2: "Profile of vaccine establishments and products currently licensed in the U.S. (1979)."
1949	Parke-Davis	Diphtheria and Tetanus Toxoids	04/08/1949		I	OTA, *A Review of Selected Federal Vaccine and Immunization Policies*, (1979), Appendix 2: "Profile of vaccine establishments and products currently licensed in the U.S. (1979)."

1949	Parke-Davis	Diphtheria Toxoid Adsorbed	04/28/1949		I	OTA, *A Review of Selected Federal Vaccine and Immunization Policies*, (1979), Appendix 2: "Profile of vaccine establishments and products currently licensed in the U.S. (1979)."
1950	Eli Lilly Laboratory	Mumps Vaccine	01/27/1950	04/11/1977	C	Inactivated.
1950	Lederle Laboratories	Mumps Vaccine	06/22/1950	05/24/1978	C	Inactivated.
1950	MPHBL	Diphtheria Toxoid Adsorbed	05/25/1950		I	OTA, *A Review of Selected Federal Vaccine and Immunization Policies*, (1979), Appendix 2: "Profile of vaccine establishments and products currently licensed in the U.S. (1979)."
1950	MPHBL	Diphtheria and Tetanus Toxoid Adsorbed	05/23/1950		I	OTA, *A Review of Selected Federal Vaccine and Immunization Policies*, (1979), Appendix 2: "Profile of vaccine establishments and products currently licensed in the U.S. (1979)."
1950	MPHBL	Tetanus Antitoxin	05/11/1950		I	
1950	MPHBL	Diphtheria and Tetanus Toxoids and Pertussis Vaccine Adsorbed	04/26/1950		I	OTA, *A Review of Selected Federal Vaccine and Immunization Policies*, (1979), Appendix 2: "Profile of vaccine establishments and products currently licensed in the U.S. (1979)."

Vaccine License Data, 1903–1999 (continued)

Year	Manufacturer	Product Name	Approved	Revoked	Innovation type	Notes
1950	BLMDPH	Diphtheria and Tetanus Toxoids Adsorbed	05/23/1950	08/27/1970	I	PHS, *Biological Products*, revised 1952.
1950	Texas Department of Health Resources	Typhoid Vaccine	07/11/1950	02/06/1979	I	
1950	University of Illinois	BCG Vaccine	07/07/1950	05/29/1987	P	A more stable version of the TICE BCG strain (originally developed at the Pasteur Institute in 1934).
1951	Bayer Corporation	Diphtheria and Tetanus Toxoids	06/01/1951	10/30/1970	I	
1951	Texas Department of Health Resources	Diphtheria and Tetanus Toxoids Adsorbed	05/11/1951	07/27/1970	I	PHS, *Biological Products*, revised 1952.
1952	Eli Lilly Laboratory	Streptococcus Vaccine	04/17/1952	10/27/1978	C	
1952	Sharp & Dhome	Cholera Vaccine	09/04/1952	01/31/1986	P	
1952	Parke-Davis	Diphtheria and Tetanus Toxoids	07/29/1952	10/14/1981	I	
1952	Parke-Davis	Diphtheria and Tetanus Toxoids and Pertussis Vaccine Adsorbed	07/29/1952	10/14/1981	I	

1952	Parke-Davis	Diphtheria Toxoid Adsorbed	02/20/1952	10/14/1981	I	
1952	Parke-Davis	Pertussis Vaccine	04/16/1952	10/14/1981	I	
1952	Parke-Davis	Pertussis Vaccine Adsorbed	02/20/1952	10/14/1981	I	
1952	Parke-Davis	Tetanus Toxoid Adsorbed	07/08/1952	10/14/1981	I	
1952	Wyeth Laboratories	Cholera Vaccine	08/16/1952		P	Agar grown; phenol-inactivated.
1952	Wyeth Laboratories	Pertussis Vaccine	07/16/1952	05/19/1987	I	
1952	Wyeth Laboratories	Typhoid Vaccine	07/16/1952		I	
1952	Wyeth Laboratories	Diphtheria and Tetanus Toxoids Adsorbed	03/07/1952		I	OTA, *A Review of Selected Federal Vaccine and Immunization Policies*, (1979), Appendix 2: "Profile of vaccine establishments and products currently licensed in the U.S. (1979)."
1953	Merrell-National	Yellow Fever Vaccine Laboratories	05/22/1953	01/03/1978	LT	This is the first recorded license, but not the original approval date. A National Drug memo indicates that this license was transferred from Rocky Mountain Labs, which began manufacturing the vaccine in 1941–1942. Original approval date may have been as early as 1935. Internal memorandum, National Drug Company, 1955, AP Archives

Vaccine License Data, 1903–1999 (continued)

Year	Manufacturer	Product Name	Approved	Revoked	Innovation type	Notes
1954	Eli Lilly Laboratory	Typhoid Vaccine	02/04/1952	11/13/1978	I	
1954	Eli Lilly Laboratory	Diphtheria and Tetanus Toxoids Adsorbed (for Adult Use)	01/14/1954		I	OTA, *A Review of Selected Federal Vaccine and Immunization Policies,* (1979), Appendix 2: "Profile of vaccine establishments and products currently licensed in the U.S. (1979)."
1954	Texas Department of Health Resources	Pertussis Vaccine	12/27/1954	02/06/1979	I	
1954	Wyeth Laboratories	Diphtheria and Tetanus Toxoids Adsorbed (for Adult Use)	12/17/1954		I	OTA, *A Review of Selected Federal Vaccine and Immunization Policies,* (1979), Appendix 2: "Profile of vaccine establishments and products currently licensed in the U.S. (1979)."
1954	Lederle Laboratories	Diphtheria and Tetanus Toxoids Adsorbed	12/17/1954	09/11/1970	I	OTA, *A Review of Selected Federal Vaccine and Immunization Policies,* (1979), Appendix 2: "Profile of vaccine establishments and products currently licensed in the U.S. (1979)."

1955	Wyeth	Tetanus Toxoids Adsorbed	06/30/1955		I	OTA, *A Review of Selected Federal Vaccine and Immunization Policies*, (1979), Appendix 2: "Profile of vaccine establishments and products currently licensed in the U.S. (1979)."
1955	Bayer Corporation	Poliovirus Vaccine Inactivated (Monkey Kidney Cell)	04/12/1955	12/28/1978	C	
1955	Merck & Company	Poliovirus Vaccine Inactivated (Monkey Kidney Cell)	04/12/1955	08/29/1980	C	
1955	Merrell-National Laboratories	Diphtheria and Tetanus Toxoids Adsorbed (for Adult Use)	03/07/1955	01/03/1978	I	
1955	Pittman Moore	Poliovirus Vaccine Inactivated	04/12/1955		C	J. Smith, *Patenting the Sun* (New York: William Morrow, 1990).
1955	Wyeth	Poliovirus Vaccine Inactivated	04/12/1955		C	J. Smith, *Patenting the Sun* (New York: William Morrow, 1990).
1955	Eli Lilly Laboratory	Poliovirus Vaccine Inactivated	04/12/1955		C	J. Smith, *Patenting the Sun* (New York: William Morrow, 1990)
1955	Parke-Davis	Poliovirus Vaccine Inactivated	04/12/1955	07/29/1980	C	

Vaccine License Data, 1903–1999 (continued)

Year	Manufacturer	Product Name	Approved	Revoked	Innovation type	Notes
1955	BLMDPH	Diphtheria Toxoid Adsorbed	08/18/1955	08/27/1970	I	OTA, *A Review of Selected Federal Vaccine and Immunization Policies*, (1979), Appendix 2: "Profile of vaccine establishments and products currently licensed in the U.S. (1979)."
1956	Bayer Corporation	Diphtheria and Tetanus Toxoids Adsorbed (for Adult Use)	11/16/1956	11/30/1970	I	
1957	Parke-Davis	Adenovirus Vaccine	09/23/1957	07/29/1980	C	Inactivated. Lederle also produced an adenovirus vaccine, although there is no record of it.
1959	Parke-Davis	Adenovirus and Influenza Virus Vaccines Combined Aluminum Phosphate Adsorbed	09/22/1959	07/29/1980	A	Inactivated.
1959	Parke-Davis	Diphtheria, Tetanus Toxoids, Pertussis, Poliomyelitis Vaccines Adsorbed	03/25/1959	10/14/1981	A	
1959	Texas Department of Health Resources	Tetanus Toxoid	09/22/1959	02/06/1979	I	

1960	Merck & Company	Poliovirus Vaccine Inactivated	07/05/1960		P	Purivax; higher purity and potency than other available IPVs. *Licensure Dates for Products Developed by Virus and Cell Biology Research* (1996), MA.
1960	Parke-Davis	Poliomyelitis Vaccine Adsorbed	10/04/1960	07/29/1980	P	
1960	Pfizer	Measles, Killed		00/00/1967	C	First measles vaccine. Killed vaccine was pulled from the market for causing atypical measles in vaccinated children who came into contact with wild measles viruses. A. Allen, *Vaccine* (New York: Norton, 2007), 231.
1960	Eli Lilly Laboratory	Measles, Killed		00/00/1967	C	A. Allen, *Vaccine* (New York: Norton, 2007), 231.
1961	Merrell-National Laboratories	Smallpox Vaccine	10/15/1961	01/03/1978	I	Freeze-dried, for government and domestic use. AP Archives.
1961	Pfizer	Poliovirus Live Oral Type 2	10/06/1961	06/12/1979	P	*The Children's Vaccine Initiative: Achieving the Vision* (Washington, DC: National Academies Press, 1993), 196–204, Appendix H: Historical Record of Vaccine Product License Holders in the U.S.

Vaccine License Data, 1903–1999 (continued)

Year	Manufacturer	Product Name	Approved	Revoked	Innovation type	Notes
1961	Pfizer	Poliovirus Live Oral Type 3	03/27/1962	06/12/1979	P	OTA, *A Review of Selected Federal Vaccine and Immunization Policies,* (1979), Appendix 2: "Profile of vaccine establishments and products currently licensed in the U.S. (1979)."
1961	Pfizer	Poliovirus Live Oral Type 1	08/17/1961	06/12/1979	P	*The Children's Vaccine Initiative: Achieving the Vision* (Washington, DC: National Academies Press, 1993), 196–204, Appendix H: Historical Record of Vaccine Product License Holders in the U.S.
1961	Wyeth	Influenza Virus Vaccine	12/13/1961		I	Flushield; split viron.
1961	Wyeth	Diphtheria, Tetanus Toxoids, Pertussis Vaccine Adsorbed	05/16/1961		I	OTA, *A Review of Selected Federal Vaccine and Immunization Policies,* (1979), Appendix 2: "Profile of vaccine establishments and products currently licensed in the U.S. (1979)."
1962	Lederle Laboratories	Diphtheria and Tetanus Toxoids Adsorbed (for Adult Use)	04/06/1962		I	OTA, *A Review of Selected Federal Vaccine and Immunization Policies,* (1979), Appendix 2: "Profile of vaccine establishments and products currently licensed in the U.S. (1979)."

1962	Lederle Laboratories	Poliovirus Live Oral Type 1	03/27/1962		I	
1962	Lederle Laboratories	Poliovirus Live Oral Type 2	03/27/1962		I	
1962	Lederle Laboratories	Poliovirus Live Oral Type 3	03/27/1962		I	OTA, *A Review of Selected Federal Vaccine and Immunization Policies*, (1979), Appendix 2: "Profile of vaccine establishments and products currently licensed in the U.S. (1979)."
1963	Lederle Laboratories	Poliovirus Live Oral Trivalent	06/25/1963	10/02/1985	A	Orimune; Sabin Strains: types 1,2,3.
1963	Merck & Company	Measles Virus Vaccine Live	03/21/1963		P	Rubeovax: live attenuated; derived from Ender's Edmonston strain and propagated in chick embryo cell cultures.
1963	Merck & Company	Typhoid Vaccine	04/25/1963	01/31/1986	I	
1963	Parke-Davis	Diphtheria, Tetanus Toxoids, Pertussis, Poliomyelitis Vaccines Adsorbed	12/20/1963	10/14/1981	I	
1965	Pfizer	Measles Virus Vaccine Live	02/00/1965	00/00/1970	P	Pfizer Vax-Measles L: www2.cdc.gov/nip/isd/immtoolkit/content/products/travelVacForeign.pdf (accessed January 16, 2011).

Vaccine License Data, 1903–1999 (continued)

Year	Manufacturer	Product Name	Approved	Revoked	Innovation type	Notes
1965	Pittman Moore	Measles Virus Vaccine Live			P	Lirugen: less reactogenic; does not require extra shot of immune gamma globulin. A. Allen, *Vaccine* (New York: Norton, 2007), 227.
1965	Dow Chemical Company	Measles Virus Vaccine Live	02/05/1965	06/21/1978	I	Cultured in dog kidney cells; withdrawn because too reactogenic.
1965	Merck & Company	Smallpox Vaccine	09/21/1965	07/29/1980	I	
1966	Lederle Laboratories	Measles Virus Vaccine Live	05/03/1966	05/21/1980	I	
1966	Pfizer	Poliovirus Live Oral Trivalent	10/28/1966	06/12/1979	I	
1967	Connaught Laboratories Inc.	BCG Vaccine	03/31/1967	05/21/1990	I	
1967	Connaught Laboratories	Smallpox Vaccine	10/23/1967	04/01/1980	I	
1967	Lederle Laboratories	Typhus Vaccine	05/24/1967	11/20/1980	P	*The Children's Vaccine Initiative: Achieving the Vision* (Washington, DC: National Academies Press, 1993), 196–204, Appendix H: Historical Record of Vaccine Product License Holders in the U.S.

1967	MPHBL	Diphtheria and Tetanus Toxoids Adsorbed (for Adult Use)	10/18/1967	07/27/1970	I	OTA, *A Review of Selected Federal Vaccine and Immunization Policies*, (1979), Appendix 2: "Profile of vaccine establishments and products currently licensed in the U.S. (1979)."
1967	MPHBL	Tetanus Toxoid Adsorbed	05/09/1967		I	OTA, *A Review of Selected Federal Vaccine and Immunization Policies*, (1979), Appendix 2: "Profile of vaccine establishments and products currently licensed in the U.S. (1979)."
1967	Merck & Company	Measles Live and Smallpox Vaccine	12/17/1967	03/12/1987	A	Combined with Dryvax from Wyeth.
1967	Merck & Company	Mumps Virus Vaccine Live	12/28/1967	01/06/1986	C	Mumpsvax; first live attenuated mumps vaccines.
1967	BLMDPH	Pertussis Vaccine Adsorbed	10/12/1967	11/12/1998	I	
1968	Bayer Corporation	Cholera Vaccine	10/03/1968	10/30/1970	P	Chemically inactivated
1968	Connaught Laboratories	Botulism Antitoxin	08/16/1968	02/24/2000	I	
1968	Merck & Company	Measles Virus Vaccine Live	11/26/1968	01/06/1986	P	Attenuvax: Fourteen additional passages of Rubeovax produced the more attenuated "Moraten" strain. *Licensure Dates for Products Developed by Virus and Cell Biology Research* (1996), MA.

Vaccine License Data, 1903–1999 (continued)

Year	Manufacturer	Product Name	Approved	Revoked	Innovation type	Notes
1969	Merck & Company	Rubella Virus Vaccine Live	06/09/1969	09/18/1978	C	Meruvax II: live attenuated virus.
1969	Phillips Roxanne Laboratories	Rubella Virus Vaccine Live	12/09/1969	01/06/1986	C	
1969	Parke-Davis	Rubella Virus Vaccine Live	00/00/1969	00/00/1972	C	Rubelogen: HPV 77 strain grown in dog kidney: www2.cdc.gov/nip/isd/immtoolkit/content/products/travelVacForeign.pdf (accessed January 16, 2011).
1970	Dow Chemical Company	Diphtheria and Tetanus Toxoids Adsorbed	09/11/1970	06/07/1977	LT	Name change. In 1970, the Division of Biologic Standards revoked product licenses and reissued new licenses for some products under the names currently used. Each instance is designated as a "Name change" in the notes section. R. Rader, *Biopharma: Biopharmaceutical Products in the U.S. Market* (Rockville, MD: Biotechnology Information Institute, 2001).
1970	Dow Chemical Company	Diphtheria and Tetanus Toxoids and Pertussis Vaccine Adsorbed	09/11/1970	06/07/1977	LT	Name change.

1970	Dow Chemical Company	Diphtheria Toxoid and Pertussis Vaccine Adsorbed	09/11/1970	06/07/1977	LT	Name change.
1970	Dow Chemical Company	Tetanus Toxoid Adsorbed	09/01/1970	06/07/1977	LT	Name change.
1970	Eli Lilly Laboratory	Diphtheria and Tetanus Toxoids	09/09/1970	06/07/1979	LT	Name change.
1970	Eli Lilly Laboratory	Diphtheria and Tetanus Toxoids Adsorbed	09/09/1970	06/07/1979	LT	Name change.
1970	Eli Lilly Laboratory	Diphtheria and Tetanus Toxoids and Pertussis Vaccine Adsorbed	09/09/1970	12/02/1985	LT	Name change.
1970	Eli Lilly Laboratory	Tetanus Toxoid Adsorbed	09/09/1970	06/07/1979	LT	Name change.
1970	Lederle Laboratories	Diphtheria and Tetanus Toxoids Adsorbed	07/29/1970		LT	Name change.
1970	Lederle Laboratories	Diphtheria and Tetanus Toxoids and Pertussis Vaccine Adsorbed	07/24/1970	08/14/1998	LT	Tri-immunol: name change.

Vaccine License Data, 1903–1999 (continued)

Year	Manufacturer	Product Name	Approved	Revoked	Innovation type	Notes
1970	Lederle Laboratories	Tetanus Toxoid Adsorbed	08/29/1970		LT	Name change.
1970	MPHBL	Diphtheria and Tetanus Toxoids Adsorbed	07/27/1970	08/03/2000	LT	Name change.
1970	MPHBL	Diphtheria and Tetanus Toxoids and Pertussis Vaccine Adsorbed	07/27/1970	12/22/1998	LT	Name change.
1970	MPHBL	Tetanus Toxoid Adsorbed	07/29/1970		LT	Name change.
1970	Merck & Company	Diphtheria and Tetanus Toxoids and Pertussis Vaccine Adsorbed	08/31/1970	01/31/1986	LT	Name change.
1970	Merck & Company	Rubella and Mumps Virus Vaccine Live	08/30/1970	01/06/1986	A	Biavax II.
1970	Merck & Company	Diphtheria and Tetanus Toxoids Adsorbed (for Adult Use)	08/31/1970	01/31/1986	LT	Name change.

1970	Merck & Company	Tetanus Toxoid Adsorbed	08/31/1970	01/31/1986	LT	Name change.
1970	Merrell-National Laboratories	Diphtheria and Tetanus Toxoids and Pertussis Vaccine	10/15/1970	01/03/1978	LT	Name change.
1970	Merrell-National Laboratories	Influenza Virus Vaccine Bivalent	04/10/1970	06/07/1979	P	Fluzone; highly purified through new ultracentrifuge process; less reactogenic. First use of automated egg harvesting. Jeff Widmer, *The Spirit of Swiftwater: 100 Years at the Pocono Labs* (Swiftwater, PA: Connaught Laboratories, 1997).
1970	Merrell-National	Tetanus Toxoid Adsorbed	10/15/1970	01/03/1978	LT	Name change.
1970	BLMDPH	Anthrax Vaccine Adsorbed	11/04/1970	11/12/1998	P	New strain and manufacturing process, less reactogenic.
1970	BLMDPH	Diphtheria and Tetanus Toxoids Adsorbed	08/27/1970	11/12/1998	LT	Name change.
1970	BLMDPH	Diphtheria and Tetanus Toxoids and Pertussis Vaccine Adsorbed	08/27/1970	11/12/1998	LT	Name change.

Vaccine License Data, 1903–1999 (continued)

Year	Manufacturer	Product Name	Approved	Revoked	Innovation type	Notes
1970	BLMDPH	Diphtheria Toxoid Adsorbed	08/27/1970	11/12/1998	LT	Name change.
1970	BLMDPH	Tetanus Toxoid Adsorbed	08/27/1970	11/12/1998	LT	Name change.
1970	SmithKline Beecham	Rubella Virus Vaccine Live	03/12/1970	10/05/1982	I	SKB also had a combination rubella-measles vaccine, but there is no record of the date it was licensed or discontinued. www2.cdc.gov/nip/isd/immtoolkit/content/products/travelVacForeign.pdf (accessed January 16, 2011).
1970	Texas Department of Health Resources	Diphtheria and Tetanus Toxoids Adsorbed	07/27/1970	02/06/1979	LT	Name change.
1970	Texas Department of Health Resources	Diphtheria and Tetanus Toxoids and Pertussis Vaccine Adsorbed	07/27/1970	02/06/1979	LT	Name change.
1970	Wyeth Laboratories	Diphtheria and Tetanus Toxoids Adsorbed	09/11/1970		LT	Name change.

1970	Wyeth Laboratories	Diphtheria and Tetanus Toxoids and Pertussis Vaccine Adsorbed	09/11/1970		LT	Name change.
1970	Wyeth Laboratories	Diphtheria Toxoid Adsorbed	09/11/1970	05/19/1987	LT	Name change.
1970	Wyeth Laboratories	Tetanus Toxoid Adsorbed	09/11/1970		LT	Name change.
1971	Merck & Company	Measles and Rubella Vaccine Live	04/22/1971	09/18/1978	A	MR Vax II.
1971	Merck & Company	Measles, Mumps, and Rubella Virus Vaccine Live	04/22/1971	09/18/1978	A	MMR Vax II.
1973	Merck & Company	Measles and Mumps Vaccine Live	07/18/1973	01/06/1986	A	MM Vax.
1974	Dow Chemical Company	Measles and Rubella Vaccine Live	04/17/1974	06/21/1978	I	
1974	Dow Chemical Company	Measles, Mumps, and Rubella Virus Vaccine Live	04/17/1974	06/21/1978	I	
1974	Dow Chemical Company	Mumps Virus Vaccine Live	04/17/1974	06/21/1978	I	
1974	Dow Chemical	Rubella Vaccine Live	03/01/1974	06/21/1978	I	

Vaccine License Data, 1903–1999 (continued)

Year	Manufacturer	Product Name	Approved	Revoked	Innovation type	Notes
1974	Merck & Company	Meningococcal Polysaccharide Vaccine, Group C	04/02/1974	03/15/1995	C	
1975	Merck & Company	Meningococcal Polysaccharide Vaccine, Group A	07/11/1975	03/15/1995	C	
1975	Merck & Company	Meningococcal Polysaccharide Vaccine, Groups A and C Combined	10/06/1975	03/15/1995	A	
1975	Merrell-National Laboratories	Meningococcal Polysaccharide Vaccine, Group A	09/19/1975	01/03/1978	C	
1975	Merrell-National Laboratories	Meningococcal Polysaccharide Vaccine, Group C	07/11/1975	01/03/1978	C	
1976	Merrell-National Laboratories	Meningococcal Polysaccharide Vaccine, Groups A and C Combined	12/13/1976	01/03/1978	A	
1977	Merck & Company	Pneumococcal Vaccine, Polyvalent	11/21/1977	01/06/1986	I	Pneumovax: added new strains: 14-valent.

1978	Connaught Laboratories	Diphtheria Antitoxin	01/03/1978	12/09/1999	LT	
1978	Connaught Laboratories	Influenza Virus Vaccine	01/03/1978	12/09/1999	LT	Fluzone.
1978	Connaught Laboratories	Tetanus Antitoxin	01/03/1978	09/22/1980	LT	From Merrell-National.
1978	Connaught Laboratories Inc.	Diphtheria and Tetanus Toxoids and Pertussis Vaccine	01/03/1978	10/29/1982	LT	From Merrell-National. Jeff Widmer, *The Spirit of Swiftwater: 100 Years at the Pocono Labs* (Swiftwater, PA: Connaught Laboratories, 1997).
1978	Connaught Laboratories Inc.	Diphtheria and Tetanus Toxoids and Pertussis Vaccine Adsorbed	01/03/1978	12/09/1999	LT	From Merrell-National.
1978	Connaught Laboratories Inc.	Diphtheria Toxoid	01/03/1978	06/21/1994	LT	
1978	Connaught Laboratories Inc.	Meningococcal Polysaccharide Vaccine, Group A	01/03/1978	12/09/1999	LT	Menomune A; from Merrell-National.
1978	Connaught Laboratories Inc.	Meningococcal Polysaccharide Vaccine, Group C	01/03/1978	12/09/1999	LT	Menomune C; from Merrell-National.

Vaccine License Data, 1903–1999 (continued)

Year	Manufacturer	Product Name	Approved	Revoked	Innovation type	Notes
1978	Connaught Laboratories Inc.	Meningococcal Polysaccharide Vaccine, Groups A and C Combined	01/03/1978	12/09/1999	LT	Menomune A/C; from Merrell-National.
1978	Connaught Laboratories Inc.	Pertussis Vaccine	01/03/1978	12/19/1997	LT	From Merrell-National.
1978	Connaught Laboratories Inc.	Smallpox Vaccine	01/03/1978	12/19/1997	LT	
1978	Connaught Laboratories Inc.	Diphtheria and Tetanus Toxoids Adsorbed (for Adult Use)	01/03/1978	12/09/1999	LT	From Merrell-National.
1978	Connaught Laboratories Inc.	Tetanus Toxoid	01/03/1978	12/09/1999	LT	
1978	Connaught Laboratories Inc.	Tetanus Toxoid Adsorbed	01/03/1978	12/09/1999	LT	From Merrell-National.
1978	Connaught Laboratories Inc.	Yellow Fever Vaccine	01/03/1978	12/09/1999	LT	YF-Vax.
1978	Merck & Company	Measles and Rubella Vaccine Live	09/18/1978	01/06/1986	I	Reformulated to include RA 27/3 strain from the Wistar Institute. *Licensure Dates for Products Developed*

						by Virus and Cell Biology Research (1996), MA.
1978	Merck & Company	Measles, Mumps and Rubella Virus Vaccine Live	09/18/1978	01/06/1986	I	MMR Vax II: reformulated to include RA 27/3 strain from the Wistar Institute. *Licensure Dates for Products Developed by Virus and Cell Biology Research* (1996), MA.
1978	Merck & Company	Rubella Virus Vaccine Live	09/18/1978		I	Replaced original strain with RA 27/3 strain developed at Wistar Institute. *Licensure Dates for Products Developed by Virus and Cell Biology Research* (1996), MA.
1979	Lederle Laboratories	Pneumococcal Vaccine, Polyvalent	08/15/1979	01/06/1986	A	Pnu-imune
1980	Wyeth Laboratories	Adenovirus Vaccine Live Oral Type 4	07/01/1980		P	Ceased manufacture in 1996.
1980	Wyeth Laboratories	Adenovirus Vaccine Live Oral Type 7	07/01/1980		P	Ceased manufacture in 1996.
1981	Connaught Laboratories Inc.	Meningococcal Polysaccharide Vaccine, Groups A,C,Y, W135 Combined	11/23/1981	12/09/1999	A	Menomune A/C/Y/W-135.
1981	Merck & Company	Hepatitis B Vaccine	11/16/1981	03/15/1995	C	

Vaccine License Data, 1903–1999 (continued)

Year	Manufacturer	Product Name	Approved	Revoked	Innovation type	Notes
1982	Merck & Company	Meningococcal Polysaccharide Vaccine, Groups A,C,Y, W135 Combined	12/14/1982	03/15/1995	A	
1982	Wyeth Laboratories	Rabies Vaccine	08/11/1982	08/07/1986	P	Non-nervous tissue origin (human diploid cells), less reactogenic.
1983	Merck & Company	Pneumococcal Vaccine, Polyvalent	07/07/1983	01/06/1986	I	Incorporated more strains for wider effectiveness: 23 valent.
1984	Connaught Laboratories Inc.	Diphtheria and Tetanus Toxoids Adsorbed	09/18/1984	12/09/1999	LT	
1985	Connaught Laboratories Inc.	Haemophilus b Polysaccharide Vaccine	12/20/1985	12/09/1999	C	HibVax.
1985	Lederle Laboratories	Haemophilus b Polysaccharide Vaccine	12/20/1985	10/23/1996	C	Hib-immune.
1985	Lederle Laboratories	Poliovirus Live Oral Trivalent	10/02/1985		LT	Name change.
1985	Praxis Biologics	Haemophilus b Polysaccharide Vaccine	04/12/1985	07/08/1994	C	b-CAPSA-1
1986	Lederle Laboratories	Pneumococcal Vaccine, Polyvalent	01/06/1986		LT	Name change.

1986	Merck & Company	Hepatitis B Vaccine (Recombinant)	07/23/1986		P	Recombivax HB; recombinant: eliminates risk of plasma derived vaccine.
1986	Merck & Company	Measles and Mumps Vaccine Live	01/06/1986		LT	Name change. *Licensure Dates for Products Developed by Virus and Cell Biology Research* (1996), MA.
1986	Merck & Company	Measles and Rubella Vaccine Live	01/06/1986		LT	Name change. *Licensure Dates for Products Developed by Virus and Cell Biology Research* (1996), MA.
1986	Merck & Company	Measles, Mumps, and Rubella Virus Vaccine Live	01/06/1986		LT	Name change. *Licensure Dates for Products Developed by Virus and Cell Biology Research* (1996), MA.
1986	Merck & Company	Mumps Virus Vaccine Live	01/06/1986		LT	Name change. *Licensure Dates for Products Developed by Virus and Cell Biology Research* (1996), MA.
1986	Merck & Company	Pneumococcal Vaccine, Polyvalent	01/06/1986		LT	Name change. *Licensure Dates for Products Developed by Virus and Cell Biology Research* (1996), MA.
1986	Merck & Company	Rubella and Mumps Virus Vaccine Live	01/06/1986		LT	Name change. *Licensure Dates for Products Developed by Virus and Cell Biology Research* (1996), MA.
1987	Bionetics Research	BCG Vaccine	05/29/1987	06/21/1989	LT	From University of Illinois

Vaccine License Data, 1903–1999 (continued)

Year	Manufacturer	Product Name	Approved	Revoked	Innovation type	Notes
1987	Connaught Laboratories Inc.	Haemophilus b Conjugate Vaccine (Diphtheria Toxoid Conjugate)	12/22/1987	12/09/1999	P	Prohibit; conjugation improves immunity; safe for children under 18 months.
1987	Connaught Laboratories Inc.	Poliovirus Vaccine Inactivated (Human Diploid Cell)	11/20/1987	02/24/2000	I	Poliovax.
1988	BLMDPH	Rabies Vaccine Adsorbed	03/18/1988	11/12/1998	I	Fetal rhesus lung culture; alum adsorbed.
1988	Praxis Biologics	Haemophilus b Conjugate Vaccine (Diphthera CRM197 Protein Conjugate)	12/21/1988	12/06/1994	I	Immunization of children 18 months to 5 years.
1989	Merck & Company	Haemophilus b Conjugate Vaccine (Meningococcal Protein Conjugate)	12/20/1989	00/00/1990	I	Pedvax Hib; immunization of children 18 months to 5 years.
1989	Organon Teknika Corporation	BCG Vaccine	06/21/1989	00/00/1994	LT	From Bionetics Research.
1989	SmithKline Beecham Biologicals	Hepatitis B Vaccine (Recombinant)	08/28/1989		I	Energix B.

1990	Connaught Laboratories Inc.	BCG Live	05/21/1990		P	Thera-Cys: new indication: treatment of carcinoma-in-situ-bladder. fda.gov/cber/efoi/approve.February 2002
1990	Merck & Company	Haemophilus b Conjugate Vaccine (Meningococcal Protein Conjugate)	00/00/1990		Is	Pedvax Hib: indication: extend use to infants 2 months and up. fda.gov/cber/efoi/approve. February 2002
1990	Praxis Biologics	Haemophilus b Conjugate Vaccine (Diphthera CRM197 Protein Conjugate)	10/04/1990		Is	Indication: use in infants. fda.gov/cber/efoi/approve. February 2002
1991	Connaught Laboratories Inc.	Rabies Vaccine	12/27/1991	02/24/2000	I	Rabie-vax: The original approval date was probably in the late 1970s although there is no record of this. R. Rader, *Biopharma: Biopharmaceutical Products in the U.S. Market* (Rockville, MD: Biotechnology Information Institute, 2001) 303.
1991	Lederle Laboratories	Diphtheria and Tetanus Toxoids Adsorbed, Acellular Pertussis Vaccine	12/17/1991	00/00/2001	A	Acel-immune: combined new aP component developed by Takeda Chemical Industry in Japan: for fourth and fifth dose of immunization schedule.

Vaccine License Data, 1903–1999 (continued)

Year	Manufacturer	Product Name	Approved	Revoked	Innovation type	Notes
1992	Connaught Laboratories Inc.	Diphtheria and Tetanus Toxoids Adsorbed, Acellular Pertussis Vaccine	08/20/1992	12/09/1999	I	Tripedia: aP component developed by Biken in Japan for fourth and fifth dose of immunization schedule.
1993	Lederle Laboratories	Diphtheria and Tetanus Toxoids and Pertussis Vaccine Adsorbed and Haemophilus b Conjugate Vaccine (Diphtheria CRM197 Protein Conjugate)	03/30/1993		A	TetraImmune.
1994	Greer Laboratories	Plague Vaccine	10/05/1994		LT	From Bayer, DOD sole purchaser, ceased manufacture in 1998 when failed GMP inspections but maintained license.
1994	Lederle Laboratories	Haemophilus b Conjugate Vaccine (Diphtheria CRM197 Protein Conjugate)	12/06/1994		LT	HibTiter: from Praxis.

1994	Organon Teknika Corporation	BCG Vaccine	00/00/1994	01/10/1995	I	New indication: treatment of bladder cancer. fda.gov/cber/efoi/approve February 2002.
1995	Merck & Company	Varicella Virus Vaccine Live	03/17/1995		C	Varivax: 12 months and older.
1995	Organon Teknika Corporation	BCG Vaccine	01/10/1995		LT	Changed manufacturing location from Chicago, IL, to Durham, NC.
1995	SmithKline Beecham Biologicals	Hepatitis A Vaccine Inactivated	02/22/1995		C	Havrix.
1996	Connaught Laboratories Inc.	Diphtheria and Tetanus Toxoids Adsorbed, Acellular Pertussis Vaccine	07/31/1996		Is	Tripedia: new indication: primary series in infant and child.
						fda.gov/cber/efoi/approve. February 2002
1996	Connaught Laboratories Inc.	Haemophilus b Conjugate Vaccine (Tetanus Toxoid Conjugate)	09/27/1996		Is	Suitable for reconstitution with DTaP. fda.gov/cber/efoi/approve. February 2002
1996	Lederle Laboratories	Diphtheria and Tetanus Toxoids Adsorbed, Acellular Pertussis Vaccine	12/30/1996		Is	Acel-immune: new indication. fda.gov/cber/efoi/approve. February 2002

Vaccine License Data, 1903–1999 (continued)

Year	Manufacturer	Product Name	Approved	Revoked	Innovation type	Notes
1996	Merck & Company	Haemophilus b Conjugate Vaccine (Meningococcal Protein Conjugate) and Hepatitis B (Recombinant) Vaccine	10/02/1996		A	Comvax: 6 weeks to 15 months; born to hepatitis B surface antigen negative mothers.
1996	Merck & Company	Hepatitis A Vaccine Inactivated	03/29/1996		P	Vaqta: new nuclease enzyme purification process.
1997	Connaught Laboratories	Diphtheria and Tetanus Toxoids Adsorbed	04/11/1997	02/24/2000	Is	Infants and children 6 weeks to 7 years.
1997	SmithKline Beecham Biologicals	Diphtheria and Tetanus Toxoids Adsorbed, Acellular Pertussis Vaccine	01/29/1997		Is	Infanrix: new indication.
1998	BioPort Corporation	Anthrax Vaccine Adsorbed	11/12/1998		LT	From Michigan Biologic Products Institute (MBPI), formerly BLMDPH.
1998	BioPort Corporation	Diphtheria and Tetanus Toxoids Adsorbed	11/12/1998		LT	From MBPI.

1998	BioPort Corporation	Diphtheria and Tetanus Toxoids and Pertussis Vaccine Adsorbed	11/12/1998		LT	From MBPI.
1998	BioPort Corporation	Diphtheria Toxoid Adsorbed	11/12/1998		LT	From MBPI.
1998	BioPort Corporation	Pertussis Vaccine Adsorbed	11/12/1998		LT	From MBPI.
1998	BioPort Corporation	Rabies Vaccine Adsorbed	11/12/1998		LT	From MBPI.
1998	BioPort Corporation	Tetanus Toxoid Adsorbed	11/12/1998		LT	From MBPI.
1998	Connaught Laboratories Inc.	BCG Vaccine	10/09/1998	02/04/2000	LT	R. Rader, *Biopharma: Biopharmaceutical Products in the U.S. Market* (Rockville, MD: Biotechnology Information Institute, 2001).
1998	North American Vaccine	Diphtheria and Tetanus Toxoids Adsorbed, Acellular Pertussis Vaccine	07/29/1998		I	Certiva: obtained exclusive aP patent from government; single purified pertussis toxoid protein detoxified with hydrogen peroxide.
1998	Parkdale Pharmaceuticals, Inc.	Influenza Virus Vaccine	04/20/1998		LT	Fluogen: from Parke-Davis, ceased manufacturing in 2000.
1998	SmithKline Beecham Biologicals	Hepatitis B Vaccine (Recombinant)	07/07/1998		Is	Energix B: indication: fda.gov/cber/efoi/approve February 2002

Vaccine License Data, 1903–1999 (continued)

Year	Manufacturer	Product Name	Approved	Revoked	Innovation type	Notes
1998	SmithKline Beecham Biologicals	Lyme Disease Vaccine (Recombinant OspA)	12/21/1998		C	Lymrix: withdrawn from market in 2002 due to poor commercial demand.
1998	Wyeth Laboratories	Rotavirus Vaccine, Live Oral Tetravalent	03/31/1998		C	Rotashield: primary immunization of infants; withdrawn from market October 15, 1999 for causing intussusception of the bowel in infants.
1999	Aventis Pasteur	Meningococcal Polysaccharide Vaccine, Group A	12/09/1999		LT	From Connaught Labs, Inc.
1999	Aventis Pasteur	Meningococcal Polysaccharide Vaccine, Group C	12/09/1999		LT	From Connaught Labs, Inc.
1999	Aventis Pasteur	Meningococcal Polysaccharide Vaccine, Groups A and C Combined	12/09/1999		LT	From Connaught Labs, Inc.
1999	Aventis Pasteur Inc.	Haemophilus b Polysaccharide Vaccine	12/09/1999		LT	From Connaught Labs, Inc.
1999	Aventis Pasteur Inc.	Yellow Fever Vaccine	12/09/1999		LT	

1999	Aventis Pasteur Limited	Diphtheria and Tetanus Toxoids Adsorbed, Acellular Pertussis Vaccine	12/09/1999	LT	Tripedia.
1999	Aventis Pasteur Limited	Diphtheria and Tetanus Toxoids and Pertussis Vaccine Adsorbed	12/09/1999	LT	
1999	Aventis Pasteur Limited	Diphtheria Antitoxin	12/09/1999	LT	
1999	Aventis Pasteur Limited	Haemophilus b Conjugate Vaccine (Diphtheria Toxoid Conjugate)	12/09/1999	LT	ActHib: from Connaught
1999	Aventis Pasteur Limited	Influenza Virus Vaccine	12/09/1999	LT	From Connaught Labs, Inc.
1999	Aventis Pasteur Limited	Meningococcal Polysaccharide Vaccine, Groups A, C, Y, W135 Combined	12/09/1999	LT	From Connaught Labs, Inc.
1999	Aventis Pasteur Limited	Diphtheria and Tetanus Toxoids Adsorbed (for Adult Use)	12/09/1999	LT	From Connaught

Vaccine License Data, 1903–1999 (continued)

Year	Manufacturer	Product Name	Approved	Revoked	Innovation type	Notes
1999	Aventis Pasteur Limited	Tetanus Toxoid	12/09/1999		LT	From Connaught Labs, Inc.
1999	Aventis Pasteur	Tetanus Toxoid Adsorbed	12/09/1999		LT	From Connaught Labs, Inc.
1999	Aventis Pasteur Limited	Pertussis Vaccine	07/29/1999		LT	From Connaught Labs, Inc.
1999	Merck & Company	Hepatitis B Vaccine (Recombinant)	08/27/1999		Is	Recombivax HB: manufacturing change; remove thimerosal preservative. fda.gov/cber/efoi/approve. February 2002
1999	Merck & Company	Hepatitis B Vaccine (Recombinant)	09/23/1999		Is	Recombivax HB: indication; two additional doses of 10mcg formulation (1.0 ml) for 0, 4, 6 months as alternative regimen for routine vaccination of adolescents 11–15 years. fda.gov/cber/efoi/approve. February 2002
1999	SmithKline Beecham Biologicals	Hepatitis B Vaccine (Recombinant)	12/14/1999		Is	Energix B: indication; fda.gov/cber/efoi/approve. February 2002

Data Sources

Unless otherwise indicated in the notes section of Appendix 1, I derived vaccine license data from the Center for Biologics Evaluation and Research of the Food and Drug Administration, *List of Vaccines Licensed in the United States since 1900*. This document was obtained under a FOIA request; file F00–16965, prepared February 28, 2001. I cross-checked all data for accuracy against primary and secondary sources. Some historical Public Health Service publications proved useful in efforts to trace the introduction of vaccines back to their original manufacturers, but these documents were published irregularly and often lacked sufficient information about the vaccine itself (i.e., whether a license was issued for an improved version of a vaccine or for a version already in production from a previous year).[1] Industry and federal archives also yielded product records and correspondence referring to company vaccine licenses.[2] I checked the accuracy of CBER's data against several anthologies and histories of vaccine development and against the less detailed information released from each agency responsible for the regulation of biologics over the past century.[3] Through these methods, I was able to correct several dozen entries on CBER's data set and to restore 99 license entries that had been lost over the years in agency shuffles. CBER's list contained 239 vaccine licenses for 1903–1999, excluding licenses issued to foreign companies and excluding licenses for vaccine ingredients produced in bulk for further manufacturing use. Using the same criteria for inclusion and exclusion, my research yielded a final list of 338 licenses for this period. While this is the most comprehensive database available, it remains incomplete. I note areas above where I have evidence that license data is missing but was unable to obtain detailed data. Many omissions remain, especially for the period prior to the 1970s. Finally, if no date is entered in the revoked box, this date was not available. It does not mean that the license was not revoked.

[1] *Biological Products: Establishments Licensed for the Preparation and Sale of Viruses, Serums, Toxins and Analogous Products, and the Trivalent Organic Arsenic Compounds* (Bethesda, MD: Public Health Service, published sporadically 1903–1937); *Establishments and Products Licensed under Section 351 of the Public Health Service Act* (Bethesda, MD, NIH [later FDA] published sporadically since 1938).

[2] Merck Archives (contains records from Mulford Labs and Sharp & Dohme), Whitehouse Station, NJ; Aventis Pasteur Archives (contains records from Pocono Labs, Swiftwater Labs, National Drug, Merrell-National, Connaught, and the Salk Institute), Swiftwater, PA; NA II (NIH and FDA records), College Park, MD; Walter Reed Army Institute of Research, records in the Joseph Smadel Reading Room, Silver Spring, MD.

[3] Stanley A. Plotkin and Edward A. Mortimer, Jr., eds., *Vaccines*, 2nd ed. (Philadelphia: W. B. Saunders, 1994); H. J. Parish, *A History of Immunization* (London: E. & S. Livingstone, 1965); Stanley A. Plotkin and B. Fantini, eds., *Vaccinia, Vaccination, and Vaccinology: Jenner, Pasteur, and Their Successors* (Paris: Elsevier, 1995); Allan Chase, *Magic Shots* (New York: William Morrow, 1982); Ronald A. Rader, *Biopharma: Biopharmaceutical Products in the U.S. Market* (Rockville, MD: Biotechnology Information Institute, 2001); Arthur Allen, *Vaccine* (New York: Norton, 2007); *Biological Products: Establishments Licensed for the Preparation and Sale of Viruses*; *Annual Reports of the Public Health and Marine Hospital Service of the United States* (Washington, DC: Treasury Department, U.S. Government Printing Office); *Establishments and Products Licensed Under Section 351 of the Public Health Service Act*; Office of Technology Assessment, *A Review of Selected Federal Vaccine and Immunization Policies: Based on Case Studies of Pneumococcal Vaccine* (Washington, DC: U.S. Government Printing Office, 1979); OTA's interpretation of CBER data provided in 1979 in Appendix 2, "Profile of vaccine establishments and products currently licensed in the U.S. (1979)."

Innovation Types

There are five categories of vaccines that represent innovative activity. These licenses are designated with a "C," "A," "P," "R," or "I."

"C" refers to "component innovations" or new vaccines that are defined as the first effective vaccine licensed to prevent a disease for which no form of active immunity was previously available.[4]

"P" refers to "process innovations," or significant improvements to the method of vaccine delivery or development. Examples include the reformulation of the adenovirus vaccine from an injected to an oral vaccine, or the transformation of the hepatitis B vaccine from a plasma-derived vaccine to a recombinant vaccine derived from yeast cells. "P" can also reflect significant enhancements to the safety or efficacy of a preexisting vaccine. It may also denote a significant new indication for a vaccine. The BCG vaccine that was licensed in 1990, for example, was approved to treat bladder cancer.

"R" refers to "radical innovations" in which a vaccine incorporates both component and process innovations. The pneumococcal vaccine developed in 1948, for example, offered the first active immunization against pneumococcal pneumonia (component innovation) using polysaccharide capsules (process innovation).

"A" refers to "architectural innovations" that combine elements of previously developed component innovations. Examples include the development of combined vaccines such as the measles, mumps, and rubella vaccine (MMR).

"I" indicates "incremental innovations" or minor improvements in the safety, efficacy, or manufacturing process. "I" also indicates vaccines that are "new" to a company, but "old" to the market. When SmithKline Beecham (SKB) followed Merck with a recombinant hepatitis B vaccine in 1989, for example, SKB's vaccine is listed as an incremental innovation, whereas Merck's 1986 vaccine is listed as an original process innovation since they were first to market. While these incremental innovations are not first to the market, they often reflect minor improvements to the safety, efficacy, or manufacturing process of the vaccine.

There are two categories of vaccines that do not represent innovative activity. In many cases, regulators issued new licenses to reflect superficial name changes or license transfers associated with industry consolidation. Similarly, in the 1990s, the FDA began to issue license supplements for trivial changes to dosing, manufacture, or indication. These licenses are designated "LT" (license transfer) and "Is" (Incremental-Supplementary), respectively.

[4] These terms (component, incremental, architectural, and radical) are derived from typologies used by Rebecca Henderson , Kim Clark, and James Utterback to distinguish among innovation types. See Rebecca Henderson and Kim B. Clark, "Architectural Innovation: The Reconfiguration of Existing Product Technologies and the Failure of Established Firms," *Administrative Science Quarterly* 35, no. 1 (March 1990): 9–30; James Utterback, *Mastering the Dynamics of Innovation* (Boston: Harvard Business School Press, 1994).

Appendix 2

Developmental History of Vaccines Licensed in the United States, 1903–1999[a]

Disease	First Isolation	Vaccine Developed	Licensure	Innovation Type[b]
Adenovirus	RI-67 1953 Hilleman, Rowe	Hilleman, Rowe: inactivated vaccine	**1957** Parke-Davis: Inactivated vaccine	Component
		1964 Chanock, Top: live vaccine[c, d]	**1980** Wyeth: live oral enteric coated	Process
Anthrax	*Bacillus anthracis:* 1876 Koch	1881 Pasteur: live attenuated vaccine for animals	1950 Merck	
		1948 Gladstone: formaldehyde inactivated, alum precipitated	1965 Michigan Department of Public Health	
			1970 Michigan Department of Public Health	Process: new strain and manufacturing process: less reactogenic
Cholera	*Vibrio cholerae:* 1854 Pacini	1896 Kolle: rudimentary agar grown, heat inactivated, whole cell vaccine	**1917** Lilly: inactivated	Incremental[e]
			1952 Merck, Wyeth	Process: chemically inactivated
Diphtheria	*Corynebacterium diphtheriae:* 1883 Klebs 1884 Loeffler toxin: 1888 Roux, Yersin	1922 Park: TAT mixtures	1922 New York Public Health	Process
		1923 Glenny, Hopkins, Gaston, Ramon: formalinized toxoid[f]	**1927** Parke-Davis diphtheria toxoid	Process: active immunization
		1928 Glenny: precipitation with alum adjuvant	**1949** Parke-Davis, Cutter Labs, Wyeth, Eli Lilly diphtheria toxoid adsorbed	Process: improved immunogenicity

Haemophilus influenzae B	1892 Pfeiffer	1920s Avery: conjugated polysaccharides and proteins to improve immunogenicity[g]	**1985** Lederle, Connaught, Praxis: unconjugated HIB vaccine	Component
		1980 Schneerson and Robbins revived work on conjugated vaccines;[h] Smith, Anderson further developed the HIB vaccine	**1987–1988** Pasteur Merieux Connaught, Praxis Biologics	Process: conjugation improves immune response by inducing proliferation of helper T-cells; safe for children under 18 months.
Hepatitis A	HM-175 strain 1976 NIAID: isolated strain and adapted host line to human diploid cells	1985 WRAIR: Inactivated[i]	**1995** SKB **1996** Merck	Component Process: first use of nuclease enzyme in purification process; improves purification without compromising yield[j]
Hepatitis B	HbsAg 1964 Blumberg[k]	Inactivated subunit vaccine; plasma derived.	**1981** Merck HBVax	Component
		Recombinant; engineer yeast cell line to express surface antigen similar to HBsAg isolated in plasma of chronic HBV carriers	**1986** Merck rHBVax[l]	Process: first recombinant vaccine; safer—carries no risk of infection associated with the use of human blood products

Developmental History of Vaccines Licensed in the United States, 1903–1999 (continued)

Disease	First Isolation	Vaccine Developed	Licensure	Innovation Type[b]
Influenza	A strain 1933 Laidlaw, Andrews, Smith[m]	1940 Francis inactivated vaccines	1942 Sharp & Dohme: inactived	Unlicensed vaccine for military use
	B strain 1940 Francis C strain 1949/1950 Francis[n]	1960–1969 Kilbourne: reassorted genes from new variants and high yield strains	**1945** Lederle and Parke-Davis: split viron; Sharp & Dohme, Lilly: whole cell	Component: lost effectiveness after 1947 due to antigenic shift
			1970 National: Zonal Centrifugation	Process: reduced reactogenicity encouraged broader use
			split and subunit purification	Process: reduced reactogenicity at expense of immunogenicity
			1971 Reassorted vaccines	Process: reduced time between variant strain identification and vaccine preparation, also required fewer eggs, limiting production costs
Japanese Encephalitis	1934 Hayashi	1940s Sabin: 10% formalinzed suspension of infected mouse brain	1945 Sharp & Dhome, Squibb & Sons	Produced unlicensed version for military use
		Lyophilized, formaldehyde inactivated Japanese	**1992** Research Foundation for Microbial Diseases	Process: more immunogenic

		encephalitis virus cultured in mouse brain[o]	of Osaka University; license also granted to Connaught/BIKEN consortium	
Lyme Disease	*Borrelia burgdorferi* 1982 Burgdorfer	Rather than induce immunity in humans, vaccine produces antibodies that are ingested by the deer tick to neutralize spirochete in the tick gut	**1998** SKB recombinant OspA	Component: (removed from the market in 2001)
Measles	Edmonston strain 1954 Enders et al.	1958 Enders, Katz, Milovanovic: passaged Edmonston strain in human tissue cultures; then adapted to propagation in chick cultures.	**1960** Pfizer, Lilly	Component: killed vaccine, revoked in 1967 for association with atypical measles
			1963 Merck	Component: live further attenuated Edmonston strain
		Schwarz: live vaccine	**1965** Pittman Moore	Process: live vaccine, less reactogenic, did not require additional shot of immune gamma globulin
		Hilleman: lyophilized live attenuated viral vaccine	**1968** Merck Attenuvax	Process: Forty additional passages created the less reactogenic Moraten strain

Developmental History of Vaccines Licensed in the United States, 1903–1999 (continued)

Disease	First Isolation	Vaccine Developed	Licensure	Innovation Type[b]
Meningococcal Meningitis	*Neisseria meningitides* 1887 Weichselbaum A and C strains 1960s Artentstein, Gotschlich, Goldschneider	1969 Artentstein, Gotschlich, Goldschneider: early polysaccharide version	**1974–1975** Merck and National: meningococcal polysaccharide vaccines A and C	Component
			1981–1982 Connaught (formerly National) and Merck: A/C/Y/W135 Combined	Incremental
Mumps	1934–1935 Johnson, Goodpasture	1948 Weller, Enders: chick embryo culture, attenuated through embryonated egg passage	**1950** Lilly and Lederle: inactivated	Component
	1963 Hilleman, Jeryl Lynn strain	1963 Hilleman: Jeryl Lynn strain passaged in chick embryo cultures	**1967** Merck: Lyophilized, live attenuated	Process: longer-lasting immunity, suitable for infants
Pertussis	*Bordetella pertussis* 1906 Bordet, Gengou	1920s Sauer: whole cell 1925 Madsen: whole cell 1942 Kendrick: alum precipitated vaccine and first combined DTP vaccine 1981 Sato: acelluar pertussis	**1914** Cutter, Lederle, Parke-Davis: whole cell pertussis vaccine	Component
			1948 Cutter: Pertussis Adsorbed	Process: more immunogenic
			1991 Lederle: combined acelluar pertussis component (Takeda	Architectural

			Chem Industry, Japan) into DTaP vaccine	
Plague	*Yersinia pestis* 1894 Kitasato and Yersin	1895 Yersin, Calmette: heat-killed, phenolized	**1942** Cutter Labs: whole cell formalin inactivated	Component: prior vaccines of unproven efficacy, Meyer reduced the dose to reduce reactogenicity
Pneumococcal Pneumonia	*Streptococcus pneumoniae* 1881 Sternberg, Pasteur Late 1960s Austrian isolated new strains and provided seed cultures for Merck vaccine	1940s Heidelberger, MacLeod: multivalent capsular polysaccharide vaccine 1960s–1970s Austrian, Hilleman	**1948** Squibb: polyvalent polysaccharide vaccine	Radical: First capsular polysaccharide vaccine
			1977 Merck: 14-valent capsular vaccine	Incremental: additional strains
			1983 Merck: 23-valent capsular vaccine	Incremental: additional strains
			2000 Lederle: pneumococcal 7-valent conjugate vaccine (Dip CRM 197 protein)[p]	Process: induces long term immunity (via T-cells) and can be used in children under 2

Developmental History of Vaccines Licensed in the United States, 1903–1999 (continued)

Disease	First Isolation	Vaccine Developed	Licensure	Innovation Type[b]
Polio	Poliomyelitis 1909 Landsteiner, Popper	1948 Koprowski et al.: OPV	**1955** Sharp & Dohme, Parke-Davis, Cutter, Pittman Moore, Wyeth, Lilly: IPV	Component: first commercially manufactured polio vaccine: inactivated, lower immunogenicity, lower reactogenicity
		1949 Enders, Robbins, Weller: propagated virus in human tissues early 1950s: Salk inactivated vaccine (IPV) grown in monkey kidney cultures[s]	**1960** Merck: IPV[q]	Incremental: higher purity and potency
		1953–1955 Sabin: OPV	**1961** Pfizer: OPV	Incremental: more immunogenic
		1954 NFIP coordinated clinical trial for Salk IPV	**1963** Lederle: OPV trivalent	Architectural
		1960 Hilleman, Charney: adjust antigenic components of Salk vaccine to improve antibody response		
Rabies	1884 Roux, Pasteur rabbit isolate	1884 Roux, Pasteur: desiccated rabbit spinal cord, saline suspension	**1915** Lilly	Incremental
	1940 Leach, Johnson human isolate	1964 Koprowski: adapted virus to human diploid cell culture[r]	**1942** Parke-Davis	Incremental
			1982 Wyeth	Process

Rocky Mountain Spotted Fever	*Rickettsia rickettsii*	1924 early version Rocky Mountain Labs (PHS) in Hamilton, MT	**1942** Lederle	Component
Rotavirus	1973 Bishop, 1974 Flewett	1984 Clark, Offit, Plotkin: first monovalent reassortment vaccine; later developed additional reassortment compounds: first grown in fetal rhesus diploid cells	**1998** Wyeth: Live, oral, tetravalent, attenuated, lyophilized	Component: withdrawn in 1999 due to increased incidence of intussusception
Rubella	1938 Hiro 1961 Parkman et al., Weller, Neva 1962 Merck Benoit strain (HPV-77)	Merck and NIH adapted virus to grow in bovine kidney cells and attenuated it through duck embryo cultures	**1969** Merck: live attenuated lyophilized	Component
	Plotkin et al: RA 27/3 strain	Cultured in human diploid lung fibroblasts	**1978** Merck: live attenuated	Incremental: less reactogenic in adults

Developmental History of Vaccines Licensed in the United States, 1903–1999 (continued)

Disease	First Isolation	Vaccine Developed	Licensure	Innovation Type[b]
Smallpox		1796 Jenner calf lymph vaccine 1968 WHO introduced quality control measures and seed lot system for global vaccine production 1973 Hekker: produced stable freeze dried vaccine in rabbit kidney cells	**1903** Pocono Labs: glycerinated calf-lymph, (in use since 1898) **1944** Lilly, Merck, Wyeth: calf-lymph	Incremental: lower bacteria counts, less reactogenic Process: lyophilization: more stable for transport
Tetanus	*Clostridium tetani* 1889 Kitasato Toxin 1890 Faber	1924 Descombey: toxoid	**1933** Parke-Davis, Sharp & Dohme: Tetanus toxoid (inactivated toxin) **1934** National: Tetanus toxoid adsorbed **1944** Wyeth	Process: active immunization Process: first recorded use of alum as adjuvant Produced unlicensed version for the military
Tuberculosis	*Mycobacterium tuberculosis* 1882 Koch	1927 Calmette, Guerin: whole cell attenuated live Tice strain (derived from *Mycobacterium*	**1950** University of Illinois: freeze-dried, live, attenuated vaccine (Tice strain) **1990** Connaught	Process: more stable, safer Process: new indication

		bovis aka *Bacille* Calmette-Guerin) 1929 Pearl: identified antitumor effect	(Toronto): licensed to use BCG to treat bladder cancer	
Typhoid	*Salmonella typhi* 1880 Erberth	1896 Pfieffer, Kolle, Wright: heat inactivated, phenol preserved	**1908** Lederle: whole cell inactive	Component: first licensure, less reactogenic
		1909 Russell: modified version used in German and British Armies[t] Whole cell heat inactivated, phenol preserved[v]	**1989** Swiss Serum and Vaccine Institute, Berna Products: live attenuated oral	Incremental: requires 4 doses, more immunogenic
		Live attenuated oral enteric coated Ty21a	**1994** Pasteur Merieux: purified Vi polysaccharide inactivated	Process: less reactogenic but short term efficacy: works less well for infants
Typhus	*Rickettsia prowazekii* 1916 Rocha Lima	1937 Zinnser: inactivated 1938 Cox: adapted to grow in chick embryo[u]	**1941** Sharp & Dohme	Process: permit large-scale production (formerly had to maintain louse farms)
	1942 Plotz isolated soluble polysaccharide antigen from rickettsial antigen	U.S. Typhus Commission developed new version of the Cox vaccine	**1967** Lederle	Process: less reactogenic, yet less immunogenic as well

Developmental History of Vaccines Licensed in the United States, 1903–1999 (continued)

Disease	First Isolation	Vaccine Developed	Licensure	Innovation Type[b]
Varicella (Chicken Pox)	Live Oka Strain 1970 Takahashi et al. Merck derived more attenuated and stable Oka/Merck strain from the Oka/BIKEN strain	Lyophilized live attenuated virus cultured in MRC-5 human diploid cells	**1995** Merck	Component
Yellow Fever	Asibi Strain and French Strain 1927	1937 Theiler, Smith: live vaccine (17D) passaged through chick embryo tissue culturex (safer than live mouse brain substrate previously used)[v]	**1953** license transferred to National Drug	Process: facilitated mass production
		1942 Hargett (Rocky Mountain Laboratory) developed a safer, aqueous base vaccine	**1987–1988** Connaught[w]	Process: safer, free of avian leukosis virus

a. This chart is not a comprehensive history, but an abstract intended to orient the reader. Multiple people, institutions, and companies contributed to the development of each vaccine. This chart represents early contributors and the first recorded licenses. Not all vaccines listed here were licensed. Vaccines that have been licensed have the date of licensure highlighted in bold type.

b. See Appendix 1 for a description of innovation types.

c. Robert M. Chanock et al., "Immunization by Selective Infection with Type 4 Adenovirus Grown in Human Diploid Tissue Culture," *JAMA* 195 (1966): 445.

d. F. H. Top et al., "Immunization with Live Types 7 and 4 Adenovirus Vaccines," *Journal of Infectious Disease* 124 (1971): 148–160.

e. Although this is the first recorded license for cholera, the Lilly vaccine is not significantly different from the original Kolle vaccine.

f. A. Glenny and B. Hopkins, "Diphtheria Toxoid as an Immunizing Agent," *British Journal of Experimental Pathology* 4 (1923): 283; in the same year, Gaston and Ramon developed a toxoid that did not need to be administered in conjunction with an antitoxin.

g. O. T. Avery and W. F. Goebel, "Chemo-Immunological Studies on Conjugated Carbohydrate Proteins," *Journal of Experimental Medicine* 50 (1929): 533.

h. R. Schneerson et al., "Preparation, Characterization, and Immunogenicity of HIB Polysaccharide-Protein Conjugates," *Journal of Experimental Medicine* 152 (1980): 361.

i. P. Provost and M. Hilleman, "Propagation of Human Hepatitis A Virus in Cell Culture in Vitro," *Proceedings of the Society of Experimental Biological Medicine* 160 (1979): 213–221.

j. Merck received the Industrial Bioprocess Award in 1998, from the American Chemical Society, for developing a process to use nuclease enzymes to purify the hepatitis A virus. This process improves purity and safety without compromising yields.

k. Baruch S. Blumberg et al., "Australia antigen and the biology of hepatitis B," *Science* 197 (1977): 17–25.

l. Merck collaborated with scientists at UCSF to clone HbsAg.

m. Wilson Smith, Christopher H. Andrewes, and Patrick P. Laidlaw, "A Virus Obtained from Influenza Patients," *Lancet* 2 (1933): 66–68.

n. Thomas Francis, Jr., et al., "Identification of Another Epidemic Respiratory Disease," *Science* 112 (1950): 495.

o. K. Takaku et al., "Japanese Encephalitis Purified Vaccine," *Biken Journal* 11 (1968): 25–39.

p. The conjugate vaccine was originally developed by Praxis, which was acquired by Lederle in 1989.

q. Merck withdrew the vaccine when they determined that it was contaminated with SV40 from monkey kidney cells.

r. T. Wiktor, M. Fernandez, and H. Koprowski, "Cultivation of Rabies Virus in Human Diploid Cell Stain WI-38," *Journal of Immunology* 93 (1964): 353–366.

s. Ernest Goodpasture developed early tissue culture techniques in the 1930s.

t. This was the first effective typhoid vaccine, according to Rose C. Engelman, *Two Hundred Years of Military Medicine* (Frederick, MD: U.S. Army Medical Department, 1975).

u. Herald R. Cox, "Use of Yolk Sac in Developing Chick Embryo as Medium for Growing Rickettsia of RMSF and Typhus Groups," *Public Health Report* 53 (1938): 2241–2247.

v. Max Theiler and Hugh H. Smith, "The Use of Yellow Fever Virus Modified by in Vitro Cultivation for Human Immunization," *Journal of Experimental Medicine* 65 (1937): 787.

w. According to a history of the National Drug Company (which had been absorbed by Connaught at this time and is now a part of Sanofi Aventis), the company licensed an improved version of the yellow fever vaccine sometime between 1987 and 1988, although I was not able to find a record of this license. Jeff Widmer, *The Spirit of Swiftwater: 100 Years at the Pocono Labs* (Swiftwater, PA: Connaught Laboratories, 1997).

Appendix 3

Twentieth Century Military Contributions to Licenses Representing Innovative Activity

Vaccine	Date	License[a]	Military Contribution
Adenovirus (inactivated vaccine)	1957	Parke-Davis	The BICIED identified the disease between 1941 and 1943. In 1953, Hilleman and Werner at WRAIR isolated the RI-67 virus from the adenoidal tissue of recruits at Fort Leonard Wood and developed a bivalent inactivated vaccine. Hilleman discovered that formalin failed to kill SV40 Simian viruses (which are oncogenic in hamsters) in the inactivated vaccine. This finding prompted the search for an alternative vaccine.
(live enteric coated oral vaccine)	1980	Wyeth	WRAIR developed a multivalent live enteric-coated oral vaccine in collaboration with NIAID: (NAID: Type 4 and WRAIR: Type 7)[b]
Anthrax (AVA)	1970	Michigan Department of Public Health	In 1948, scientists at Camp Detrick, in collaboration with Gladstone at the Lister Institute, developed the first pilot lot. Camp Detrick sponsored field trials of the vaccine that Merck manufactured for the military in the 1950s. USAMRIID conducted animal efficacy and cohort safety studies in the 1990s.
Cholera	1952	Merck, Wyeth	In the 1940s, the AMS selected and standardized the strains (Inaba and Ogawa serotypes) used in commercial vaccine production. NMRC tested new vaccines in Panama and Peru in the 1990s.[c]
Diphtheria (toxoid adsorbed)	1949	Parke-Davis, Cutter Labs, Wyeth, Eli Lilly	Army studies indicated that smaller doses of toxoid reduced reactogenicity without compromising immunogenicity. This finding permitted the development of combined diphtheria tetanus toxoids with fewer adverse effects.[d]

Hepatitis A	1995 1996	SmithKline Beecham Merck	WRAIR scientists developed the first pilot lot in 1985 and demonstrated human immunogenicity in 1986. WRAIR transferred the manufacturing capability to SmithKline Beecham the following year, training corporate scientists and assisting with scale-up pilot procedures under a no-dollar agreement. In 1988, AFRIMS and the Thai Ministry of Health conducted large-scale clinical trials in Thailand. USAMMDA coordinated the FDA licensure application for SmithKline Beecham. Merck built on WRAIR's work to license a second vaccine the following year. MRMC provided Merck with technical assistance for the clinical trials as well.[e]
Hepatitis B	1981	Merck	From 1944–1970s AFEB sponsored etiological studies at Yale University and the Willowbrook State School.[f] WRAIR determined HBsAg subtypes by geographic region.[g]
Influenza (whole cell and split viron)	1945	Sharp & Dohme Lederle, Parke-Davis, Lilly	The Influenza Commission developed and tested the first inactivated vaccine during World War II. Sharples centrifuge purification techniques were developed under CMR contracts. In 1947, WRAIR scientists determined the need for annual adjustments to the antigenic content of the influenza vaccine. In 1957, Hilleman and Buescher at WRAIR isolated and characterized the Asian influenza (H2N2) strain used in commercial vaccines.
(split and subunit)	1970	National Drug	In the 1960s, a former WRAIR scientist transferred zonal centrifugation technology from military labs to National Drug. This technology permitted the manufacture of less reactogenic influenza vaccines.

Twentieth Century Military Contributions to Licenses Representing Innovative Activity (continued)

Vaccine	Date	License[a]	Military Contribution
Japanese Encephalitis	1992	Research Foundation for Microbial Diseases of Osaka University; Connaught/BIKEN consortium	Sabin developed the first vaccine under the BICIED during World War II. The army transferred the technology to Japan during post–World War II occupation. WRAIR scientists conducted vector studies between 1952 and 1955. In the 1980s, WRAIR conducted clinical trials with the Thai Ministry of Health demonstrating that the BIKEN vaccine was 91 percent effective.[h] In the 1990s, WRAIR sponsored research in Korea to develop a new vaccine derived from WRAIR's formalin inactivated cell culture prototype.[i] Intercell licensed this technology from the army and obtained FDA approval in 2009.
Measles	1960	Pfizer, Lilly	John Enders developed the first vaccine at Harvard Medical School working under an army contract.[j] The AFEB Commission on Viral Infections sponsored clinical trials of live and inactivated vaccines to determine safety and efficacy.
Meningococcal Meningitis	1974 1975	Merck National Drug	WRAIR scientists isolated the Group A and Group C strains, developed pilot lots of meningococcal polysaccharide vaccines A and C, and conducted clinical trials in the late 1960s and 1970s.
Plague	1942	Cutter Labs	OSRD sponsored Karl Meyer's vaccine research at Hooper foundation UCSF during World War II. Earlier versions of this vaccine were of unproven efficacy. Meyer reduced the dose to reduce reactogenicity. Military use of this vaccine during the Vietnam war provided further evidence of efficacy.
Pneumococcal Pneumonia	1948	Squibb	In 1881, George Sternberg at the AMS isolated a pneumococcus bacterium. In 1945, the AMS performed the first large-scale, randomized field trial to demonstrate the efficacy of the polyvalent polysaccharide vaccine.[k]

Rubella	1969	Merck	WRAIR scientists (Parkman, Buescher, and Artenstein) isolated the rubella virus in collaboration with scientists at the Harvard School of Public Health.[l] Merck used isolation and virology techniques developed at WRAIR to develop the commercial vaccine.
Smallpox	1944	Lilly, Merck, Wyeth	The U.S. Army adapted the Jenner cowpox vaccine to military use in 1882. Sternberg of the AMS demonstrated that antibodies in the bloodstream neutralized the vaccinia virus.[m] In the 1940s, the Army Blood Program developed lyophilization techniques. In 1949, the AFEB sponsored studies to apply this technique to stabilize the vaccine for shipment.[n]
Tetanus	1944	Wyeth	Routine military use of the vaccine during World War II provided informal proof of the efficacy of the vaccine, paving the way for commercial use after the war.[o] In 1944, the American Academy of Pediatrics recommended routine administration of the vaccine to children. Research conducted by the AFEB Commission on Immunization (directed by J. Smadel) and the CMR (Mueller under contract at HMS) reduced the reactinogenicity of the vaccine by synthesizing an alternative to peptones originally found in broth cultures. After the war, the AFEB determined that the alum-adsorbed vaccine used by the navy was superior to the nonadsorbed vaccine used by the army.
Typhoid	1908 1914	Lederle Wyeth	In 1898, Army Typhoid Board members identified the vector (human contact and flies) and Major Walter Reed determined the pathogenesis of the disease. In 1909, Army Captain Frederic Russell modified a version of the vaccine used in German and British armies and conducted the first clinical studies at the AMS.[p] Widespread use of this vaccine during World War I provided informal proof of efficacy.[q] The DOD sponsored studies in Chile demonstrating 67 percent efficacy of an oral enteric-coated live attenuated Ty21a vaccine. The Naval Medical Research Unit in Jakarta (NAMRU-2) demonstrated 42 percent efficacy.

Twentieth Century Military Contributions to Licenses Representing Innovative Activity (continued)

Vaccine	Date	License[a]	Military Contribution
Typhus	1989 1941	Berna Sharp & Dohme	The AMS demonstrated that the original Cox formulation had lost potency and created the U.S. Typhus Commission to develop a new vaccine. In 1940, the commission isolated, purified, grew, produced, and tested a new typhus vaccine. This researched pooled U.S.- and U.K.-based research and epidemiological information.[r] The CMR and AMS assisted the commission by developing fractionation procedures to isolate antigens from rickettsial bodies. WRAIR studies later determined that this purified vaccine was of insufficient potency.
Yellow Fever	1953	The Rockefeller Foundation and the Rocky Mountain Laboratory supplemented vaccine production for the U.S. military during World War II.	1900–1902 Yellow Fever Commission (directed by Walter Reed) demonstrated transmission via *Aedes aegypti* mosquito and inferred the presence of filterable virus. Hepatitis B–contaminated serum used to stabilize the vaccine caused fifty thousand cases of jaundice and sixty-two deaths.[s] This event forced researchers to devise alternatives to serum-based vaccines.

a. License holders shown here are not representative of all manufacturers holding a license to this vaccine or later versions of it. For a more comprehensive list of all vaccines and license holders, see Appendix 1.

b. Robert M. Chanock et al., "Immunization by Selective Infection with Type 4 Adenovirus Grown in Human Diploid Tissue Culture," *JAMA* 195 (1966): 445; F. H. Top et al., "Immunization with Live Types 7 and 4 Adenovirus Vaccines," *Journal of Infectious Disease* 124 (1971): 148–160.

c. D. N. Taylor et al., "Expanded Safety and Immunogenicity of a Bi-Valent Oral Attenuated Cholera Vaccine, CVD 103-HgR Plus CVD 111 in United States Military Personnel Stationed in Panama," *Infection and Immunity* 67 (1999): 2030–2034; D. N. Taylor et al., "Two-Year Study of the Protective Efficacy of the Oral Whole Cell Plus Recombinant Sub-Unit Cholera Vaccine in Peru," *Journal of Infectious Diseases* 181 (2000): 1667–1673.

d. Geoffrey Edsall, James S. Altman, and Andrew J. Gaspar, "Combined Tetanus-Diphtheria Immunization for Adults," *American Journal of Public Health* 44 (1954): 1537.

e. Communication with Charles Hoke, December 30, 2010.

f. Saul Krugman and Robert McCollum (NYU) conducted a series of controversial tests when they injected live virus preparations to measure the effectiveness of immunoglobulins in patients at the Willowbrook School in Staten Island, NY.

g. W. H. Bancroft, F. K. Mundon, P. K. Russell, "Detection of Additional antigenic determinants of hepatitis B antigen," *Journal of Immunology*, 1972; 109(4):842–848.

h. Charles Hoke et al., "Protection against Japanese Encephalitis by Inactivated Vaccines," *New England Journal of Medicine* 319 (1988): 608–614.

i. Communication with Charles Hoke, December 30, 2010.

j. Preventive Medicine Research Division of the U.S. Army, Annual Report, FY '63, August 27, 1963, NA: RG 112, E. 1015, B. 106.

k. Colin M. MacLeod et al., "Prevention of Pneumococcal Pneumonia by Immunization with Specific Capsular Polysaccharides," *Journal of Experimental Medicine* 82 (1945): 445.

l. P. Parkman, E. Buescher, and M. Artenstein, "Recovery of Rubella Virus from Army Recruits," *Proceedings of the Society of Experimental Biological Medicine* 111 (1962): 225–230.

m. Rose C. Engelman, *Two Hundred Years of Military Medicine* (Frederick, MD: U.S. Army Medical Department, 1975).

n. Minutes, 3rd Annual Meeting of the Commission on Immunization, March 21, 1949, Armed Forces Epidemiology Board, NA: RG 112, E. 1035, B. 83.

o. Only twelve cases of tetanus were reported in the U.S. Army between 1939 and 1945, six of which were soldiers that had not been actively immunized. H. J. Parish, *A History of Immunization* (London: E. and S. Livingstone, 1965).

p. Russell's whole-cell, heat-inactivated, phenol-preserved version was the first effective typhoid vaccine, according to Engelman, *Two Hundred Years of Military Medicine* (Frederick, MD: U.S. Army Medical Department, 1975).

q. During the Spanish-American War of 1898, the United States suffered 20,738 cases of typhoid fever. Following routine administration of the vaccine in 1911, the United States suffered only 1,529 cases of the disease. Ronald A. Rader, *Biopharma: Biopharmaceutical Products in the U.S. Market* (Rockville, MD: Biotechnology Information Institute, 2001), 265.

r. Harry Plotz, *Report on Contributions from the Division of Virus and Rickettsial Disease, Army Medical School, on the Development of the Typhus Vaccine* (Washington, DC: Army Medical School, 1946), NA: RG 112, E. 1035, B. 83. According to Plotz, "It would be difficult to state with any degree of accuracy when each worker made a particular contribution to the development of the vaccine. This is due to the more or less constant exchange of opinions between the workers in the different laboratories. It would appear that the three laboratories actively engaged in the development of the vaccine were the National Institute of Health, the Connaught Laboratories and the Army Medical School . . . the National Institute of Health workers as well as the workers in this laboratory independently demonstrated the existence of a soluble antigen at the same time" (12).

s. Subsequent studies demonstrated that as many as 330,000 may have been infected. L. B. Seeff, "A serologic follow-up of the 1942 epidemic of post-vaccination hepatitis in the United States Army," *New England Journal of Medicine* 316, no. 16 (April 16, 1987): 965–970.

Notes

Introduction

1. Ralph H. Major, *Fatal Partners: War and Disease* (New York: Doubleday, Doran, 1941); Paul Steiner, *Disease in the Civil War: Natural Biological Warfare in 1861–1865* (Springfield, IL: Charles C. Thomas Pub Ltd., 1968); Ann M. Becker, "Smallpox in Washington's Army: Strategic Implications of the Disease during the American Revolutionary War," *Journal of Military History* 68 (April 2004): 381–430.

2. Alfred W. Crosby, *Ecological Imperialism: The Biological Expansion of Europe, 900–1900* (Cambridge: Cambridge University Press, 1986).

3. Mark Wheelis, "Biological Warfare before 1914," in *Biological and Toxin Weapons: Research Development and Use from the Middle Ages to 1945*, ed. Erhard Geissler and John Ellis van Courtland Moon, SIPRI Chemical and Biological Warfare Studies (Oxford: Oxford University Press, 1999); Vincent J. Derbes, "De Mussi and the Great Plague of 1348: A Forgotten Episode of Bacteriological War," *JAMA* 196, no. 1 (1966): 59–62.

4. Mark Wheelis, "Biological Warfare at the 1346 Siege of Caffa," *Emerging Infectious Disease* 8, no. 9 (2002): 971–975.

5. U.S. Department of State, *Adherence to and Compliance with Arms Control, Nonproliferation, Disarmament Agreement and Commitments* (Washington, DC: Bureau of Verification and Compliance, U.S. Department of State, 2005), 17–31. Iraq, Libya, South Africa, and Rhodesia invested in biological weapons programs as well, but these have since been terminated. See Gregory D. Koblentz, *Living Weapons: Biological Warfare and International Security* (Ithaca, NY: Cornell University Press, 2009), 18n28.

6. Alfred D. Chandler, Jr., *Scale and Scope: The Dynamics of Industrial Capitalism* (Cambridge, MA: Harvard University Press, 1990); Leonard S. Reich, *The Making of American Industrial Research: Science and Business at GE and*

Bell, 1876–1926 (Cambridge: Cambridge University Press, 1985); David A. Hounshell and John Kenly Smith, *Science and Corporate Strategy: Du Pont R & D, 1902–1980* (New York: Cambridge University Press, 1988); Emerson W. Pugh, *Building IBM: Shaping an Industry and Its Technology* (Cambridge, MA: MIT Press, 1984); Giovanni Dosi, "Technological Paradigms and Technological Trajectories: A Suggested Interpretation of the Determinants and Directions of Technological Change," *Research Policy* 11 (1982): 147–162; R. Foster, "Timing Technological Transitions," in *Technology in the Modern Corporation: A Strategic Perspective*, ed. Mel Horwitch (New York: Pergamon, 1986).

7. Louis Galambos and Jane Eliot Sewell, *Networks of Innovation: Vaccine Development at Merck, Sharp & Dohme, and Mulford, 1895–1995* (Cambridge: Cambridge University Press, 1995); Office of Technology Assessment, *A Review of Selected Federal Vaccine and Immunization Policies: Based on Case Studies of Pneumococcal Vaccine* (Washington, DC: U.S. Government Printing Office 1979); Institute of Medicine, *Vaccine Supply and Innovation* (Washington, DC: National Academies Press, 1985); Henry G. Grabowski and John M. Vernon, *The Search for New Vaccines* (Washington, DC: AEI Press, 1997).

8. Michael Kremer and Rachel Glennerster, *Strong Medicine: Creating Incentives for Pharmaceutical Research on Neglected Diseases* (Princeton, NJ: Princeton University Press, 2004).

9. Jean Lave and Etienne Wenger, *Situated Learning; Legitimate Peripheral Participation*, Cambridge University Press, 1991.

10. Etienne Wenger, Richard McDermott, and William Snyder, *Cultivating Communities of Practice: a Guide to Managing Knowledge*, Cambridge: Harvard Business School Press, 2002

11. Eric Von Hippel, " 'Sticky Information' and the Locus of Problem Solving: Implications for Innovation," *Management Science*, Vol. 40, No. 4 (April 1994), 429.

12. Leonard S. Reich, *The Making of American Industrial Research: Science and Business at GE and Bell, 1876–1926* (Cambridge: Cambridge University Press, 1985); Steven Shapin, *The Scientific Life: A Moral History of a Late Modern Vocation* (Chicago: University of Chicago Press, 2008), 166–167. According to historian Steven Shapin, "The contract system binding academic research to the Federal government, industry sponsorship of science, the scale and expense of scientific instruments, the specialization of scientific knowledge and the division scientific labor, and, above all, the planning and organization of scientific research as collective work that flowed from these other conditions—they were all new: the offspring of the Manhattan Project, the MIT Radiation Laboratory, the Johns Hopkins Applied Physics Laboratory, the postwar weapons laboratories, and the teamwork of particle physics."

13. Merritt Roe Smith, *Harper's Ferry Armory and the New Technology: The Challenge of Change* (Ithaca, NY: Cornell University Press, 1977); David A. Mindell, *Between Human and Machine: Feedback, Control, and Computing before Cybernetics* (Baltimore: Johns Hopkins University Press, 2002).

14. The military made significant contributions to vaccines for eighteen out of the twenty-eight vaccine-preventable diseases. See Appendices 2 and 3.

15. Eric von Hippel, Stefan Thomke, and Mary Sonnack, "Creating Breakthroughs at 3M," *Harvard Business Review*, September/October 1999, pp. 3–9.

1. Disease, Security, and Vaccines

1. The Soviet Union prepared its first report on biological weapons in 1928 and operated a small offensive program since World War II. It expanded its offensive program under Biopreparat in 1972. Soviet activities under this program dwarfed U.S. activities which were terminated in 1969. The USSR employed sixty-five thousand personnel in BW programs, whereas the United States employed fewer than six thousand; the USSR operated sixty facilities, whereas the United States operated only three; the USSR standardized fifty-two agents, whereas the United States standardized eleven; and the USSR produced five thousand tons of anthrax annually, whereas the United States produced only nine tons. Gregory Koblentz, "Proliferation of Biological Weapons" (MIT BW Workshop, Cambridge, MA, May 7, 2001).

2. D. A. Henderson to P. Jahrling, [1998] recorded in D. A. Henderson, "Bioterrorism as a Public Health Threat," *Emerging Infectious Diseases* 4, no. 3 (1998) 488–492.

3. For a more comprehensive list of bioterrorism incidents prior to 2001, see W. Seth Carus, "Bioterrorism and Biocrimes: The Illicit Use of Biological Agents since 1900" (working paper, National Defense University, August 1998 [February 2001 revision]).

4. For reasons that are not fully understood, no one died from these attempts. It is possible that the cult grew weak microbial strains (due to inappropiate strain selection, poor culture conditions, and/or contamination) or that they did not know how to aerosolize these agents appropriately. Jessica Stern, *The Ultimate Terrorists* (Cambridge, MA: Harvard University Press, 1999), 64. Richard J. Danzig et al., "Aum Shinrikyo: Insights into How Terrorists Develop Biological and Chemical Weapons," Center for a New American Security, July 2011, www.cnas.org/node/6703 (accessed July 2011).

5. Judith Miller, Stephen Engelberg, and William Broad, *Germs: Biological Weapons and America's Secret War* (New York: Simon and Schuster, 2002).

6. Jonathan B. Tucker, "Historical Trends Related to Bioterrorism: An Empirical Analysis," *Emerging Infectious Diseases* 5 (1999): 4. See also Jonathan B. Tucker, ed., *Toxic Terror: Assessing Terrorist Use of Chemical and Biological Weapons* (Cambridge, MA: MIT Press, 2000).

7. J. S. Nye, Jr., and R. J. Woolsey, "Heed the Nuclear, Biological, and Chemical Terrorist Threat," *International Herald Tribune*, June 5, 1997.

8. Bruce Hoffman, "Terrorism and WMD: Some Preliminary Hypotheses," *Nonproliferation Review* 4 (1997): 3.

9. Charles L. Mercier, Jr., "Terrorists, WMD, and the U.S. Army Reserve," *Parameters* 27 (1997): 98–118.

10. Committee on International Science, Engineering, and Technology, *Infectious Disease: A Global Health Threat* (Washington, DC: National Science and Technology Council, 1995).

11. Institute of Medicine, *Emerging Infections: Microbial Threats to Health in the United States*, ed. Joshua Lederberg, Robert E. Shope, and Stanley C. Oaks (Washington, DC: National Academies Press, 1992).

12. David F. Gordon, Don Noah, and George Fidas, "The Global Infectious Disease Threat and Its Implications for the United States," *National Intelligence Estimate* 99-17D (January 2000): 122.

13. Ibid.

14. Institute of Medicine, *Emerging Infections.*

15. www.cdc.gov/malaria/faq.htm (accessed February 2, 2010).

16. Gordon, Noah, and Fidas, "Global Infectious Disease Threat," 23.

17. Samuel Berger, "Building on the Clinton Record," *Foreign Affairs* 79, no. 6 (September/October 2000): 32.

18. Kenneth Bernard, quoted in William Muraskin, ed., *Vaccines for Developing Economies: Who Will Pay?* (New Canaan, CT: Albert B. Sabin Vaccine Institute, 2001), 79.

19. Gordon, Noah, and Fidas, "Global Infectious Disease Threat."

20. George W. Bush, *The National Security Strategy of the United States of America* (Washington, DC: White House, 2006), 44.

21. The National Security Council also made this connection a month earlier in the National Strategy for Countering Biological Threats, November 2009, www.whitehouse.gov/sites/default/files/National_Strategy_for_Countering_BioThreats.pdf (accessed November 30, 2010).

22. Department of Health and Human Services, Office of the Assistant Secretary for Preparedness and Response, *The National Health Security Strategy of the United States of America*, December 2009), 3, www.hhs.gov/aspr/opsp/nhss/nhss0912.pdf (accessed February 10, 2010).

23. Due in part to its new status as a security threat, international spending on HIV/AIDS initiatives rose from $1.5 billion in 2000 to $10 billion in 2007. Joint UN Program on HIV/AIDS, "Financial Resources Required to Achieve Universal Access to HIV Prevention, Treatment, Care, and Support" (Geneva: UNAIDS, 2007), 4. Quoted in Gregory Koblentz, "Biosecurity Reconsidered: Calibrating Biological Threats and Responses," *International Security* 34, no. 4 (Spring 2010): 96–132.

24. Motor vehicle accidents are the leading cause of death for all age groups under forty-four. They are the third-leading cause of death after cancer and heart disease for ages forty-four to sixty-five. www.disastercenter.com/cdc (accessed September 8, 2010).

25. Richard Danzig, *Catastrophic Bioterrorism: Toward a Long-Term Strategy for Limiting the Risk* (Washington, DC: National Defense University, Center for Technology and National Security Policy, 2007). See also A. Carter, J. Deutch, and P. Zelikow, "Catastrophic Terrorism: Tackling the New Danger," *Foreign Affairs* 77, no. 6 (November/December 1998): 80–94.

26. Koblentz, "Biosecurity Reconsidered," 96–132.

27. Bruce Hoffman, *Inside Terrorism* (New York: Columbia University Press, 2006), 36.

28. William H. McNeill, *Plagues and Peoples* (New York: Doubleday, 1976); Hans Zinsser, *Rats, Lice and History* (Boston: Little, Brown, 1934); Ralph H. Major, *Fatal Partners: War and Disease* (New York: Doubleday, Doran, 1941); Paul E. Steiner, *Disease in the Civil War: Natural Biological Warfare in 1861–1865*

(Springfield, IL: Charles C. Thomas Pub Ltd., 1968); Alfred W. Crosby, Jr., *Epidemics and Peace, 1918* (Westport, CT: Greenwood, 1976).

29. Andrew Price-Smith, *Contagion and Chaos: Disease, Ecology, and National Security in the Era of Globalization* (Cambridge, MA: MIT Press, 2009). Other exceptions include P. W. Singer, "AIDS and International Security," *Survival* 44, no. 1 (2002): 145–158; S. Peterson, "Epidemic Disease and National Security," *Security Studies* 12, no. 2 (2002–2003): 43–80; Y. Huang, *Mortal Peril: Public Health in China and Its Security Implications*, Health and Security Series Special Report no. 7 (Washington, DC: Chemical and Biological Arms Control Institute, 2003).

30. Price-Smith, *Contagion and Chaos*, 96.

31. Ibid., 7.

32. Variolation preceded the use of Jenner's smallpox vaccine, which was not developed until 1796. The practice of variolation introduced dried pus from smallpox pustules through a break in the skin. An estimated 2 percent to 3 percent died from full-blown cases of variolation-induced smallpox. See Elizabeth A. Fenn, *Pox Americana: The Great North American Smallpox Epidemic* (New York: Hill and Wang, 2001).

33. Ann M. Becker, "Smallpox in Washington's Army: Strategic Implications of the Disease during the American Revolutionary War," *Journal of Military History* 68 (April 2004): 381–430.

34. While effective methods of disease control conferred tremendous advantages, then as now, these advantages were not always widely recognized or consciously exploited. Demographer Philip Curtin demonstrates that European advances in hygiene and tropical medicine permitted the widespread colonization of tropical regions in the nineteenth century. Superior methods of disease control were instrumental to their success, but these methods were not part of an explicit strategy. Price-Smith notes that the British were an exception to this rule; they developed "imperial medicine" precisely for the purpose of projecting power into epidemiologically hostile regions (e.g., South Asia, Africa). Philip D. Curtin, *Death by Migration: Europe's Encounter with the Tropical World in the Nineteenth Century* (Cambridge: Cambridge University Press, 1989); Price-Smith, *Contagion and Chaos*, 195.

35. The White House, The National Strategy to Combat Weapons of Mass Destruction, December 2002, www.whitehouse.gov/news/releases/2002/12/WMDStrategy.pdf. While the 2009 National Strategy for Combating Biological Threats is more sensitive to the role of public health and conduct in the life sciences, this strategy still relies on risk prediction (which is, and will likely remain, poor) and the use of law enforcement, treaties, and "international engagement" to disrupt, interdict, and apprehend perpetrators. www.whitehouse.gov/sites/default/files/National_Strategy_for_Countering_BioThreats.pdf.

36. Mark Wheelis and Malcolm Dando, "On the Brink: Biodefence, Biotechnology, and the Future of Weapons Control," *CBW Conventions Bulletin*, no. 58 (December 2002): 3; Jonathan B. Tucker, "The Proliferation of Chemical and Biological Weapons Materials and Technologies to State and Sub-State Actors," statement, November 7, 2001, Before the

Subcommittee on International Security, Proliferation, and Federal Services of the U.S. Senate Committee on Governmental Affairs, 107th Cong., 1st Sess.

37. For a description of the agreement, see www.australiagroup.net.

38. Tucker, "Proliferation of Chemical and Biological Weapons Materials and Technologies"; J. Steinbruner et al., "Controlling Dangerous Pathogens," www.cissm.umd.edu/papers/files/pathogens_project_monograph.pdf (accessed June 14, 2011).

39. Matthew Meselson, "Averting the Hostile Exploitation of Biotechnology," *CBW Conventions Bulletin*, no. 48 (June 2000): 16–19.

40. Gregory Koblentz, *Living Weapons: Biological Warfare and International Security* (Ithaca, NY: Cornell University Press, 2009), 141–199.

41. The White House, National Strategy to Combat Weapons of Mass Destruction.

42. See the discussion in Koblentz, *Living Weapons*, 21–32.

43. Philip K. Russell, "Vaccines in Civilian Defense against Bioterrorism," *Emerging Infectious Diseases* 5, no. 4 (July/August 1999): 532.

44. Richard Danzig, *Catastrophic Bioterrorism: What Is to Be Done?* (Washington, DC: National Defense University, Center for Technology and National Security Policy, August 2003).

45. National Research Council, *Making the Nation Safer: The Role of Science and Technology in Countering Terrorism* (Washington, DC: National Academies Press, 2002), 87. New research could produce exceptions to this rule in the near future. Antisense drugs, for example, contain synthesized DNA or RNA strands that bind to key portions of a virus to prevent it from replicating. While some early results have been promising, these drugs remain largely experimental.

46. www.cdc.gov/od/sap/docs/salist.pdf.

47. Congressional Budget Office Cost Estimate: S. 15 Project Bioshield Act of 2003, May 7, 2003, 9.

48. See Chapter 4 for details on the diplomatic use of vaccines during the Cold War.

49. Animal vaccines can serve this purpose as well, while contributing to global food security. David N. Fishman and Kevin B. Laupland, "The 'One Health' Paradigm: Time for Infectious Diseases Clinicians to Take Note?" *Canadian Journal of Infectious Diseases and Medical Microbiology* 21, no. 3 (Autumn 2010): 111–114; Wheelis and Dando, "On the Brink." (I am indebted to David Franz for this insight.)

50. Other medical countermeasures, such as antivirals and broad-spectrum antibiotics, would also be useful in efforts to deter or mitigate biological attacks.

51. Justin Gillis, "Scientists Race for Vaccines: Drug Companies Called Key to Bioterror Fight," *Washington Post*, November 8, 2001.

52. Category A agents pose the highest threat to national security because they (1) can be easily disseminated or transmitted from person to person; (2) they have a high mortality rate; (3) they are likely to cause high levels of public panic; and/or (4) they require special action for public health preparedness. www.bt.cdc.gov/agent/agentlist_category.asp#adef.

53. L. M. Joellenbeck et al., eds., *The Anthrax Vaccine: Is It Safe? Does It Work?* (Washington, DC: National Academies Press, April 2002), www.nap.edu/catalog/10310.html.

54. To overcome this ethical obstacle, the FDA adopted the Animal Efficacy Rule in 2002, which allows biodefense vaccines to be licensed on the basis of animal studies alone.

55. *Ensuring Biologics Advanced Development and Manufacturing Capability for the United States Government: A Summary of Key Findings and Conclusions*, Cooperative Agreement Research Study between Defense Applied Research Projects Agency (DARPA) and University of Pittsburgh Medical Center (UPMC), July 2007–March 2009, http://oai.dtic.mil/oai/oai?verb=getRecord&metadataPrefix=html&identifier=ADA506569 (accessed November 28, 2010).

56. Ibid.

57. J. Matheny, M. Mair, and B. Smith, "Cost/Success Projections for U.S. Biodefense Countermeasure Development," *Nature Biotechnology* 26, no. 9 (September): 981–983, available at:www.nature.com/nbt/journal/v26/n9/full/nbt0908–981.html. Subsequent studies have used more conservative measures, estimating the actual cost to be between $6.3 billion and $11.6 billion for 2009–2015. L. Klotz and A. Pearson, "BARDA's Budget," *Nature Biotechnology* 27, no. 8 (August 2009): 698.

58. For an account of this operation, code named BACUS (biotechnology activity characterization by unconventional signatures), see Miller, Engelberg, and Broad, *Germs*.

59. National Academy of Sciences, *Globalization, Biosecurity, and Future of the Life Sciences* (Washington, DC: National Academies Press, 2006).

60. Iraq did have anthrax in the first Gulf War but failed to use it.

61. *Ensuring Biologics Advanced Development and Manufacturing Capability for the United States Government*, DARPA and UPMC, www.dtic.mil/cgi-bin/GetTRDoc?AD=ADA506569&Location=U2&doc=GetTRDoc.pdf (accessed February 2010); R. Danzig and K. Hoyt, *Recommendations for a DOD Vaccine Program* (DARPA Special Projects Office, 2005). A number of DOD and HHS programs support this strategy. Examples include DARPA's Transformational Medical Technologies (TMT) and Alternative Manufacturing Process (AMP) programs.

62. U.S. Department of Health and Human Services, The Public Health Emergency Medical Countermeasures Enterprise Review: Transforming the Enterprise to Meet Long-Range National Needs, August 2010, www.phe.gov/Preparedness/mcm/enterprisereview/Pages/default.aspx (accessed November 23, 2010).

2. Historical Patterns of Vaccine Innovation

1. Office of Technology Assessment, *A Review of Selected Federal Vaccine and Immunization Policies: Based on Case Studies of Pneumococcal Vaccine* (Washington, DC: U.S. Government Printing Office, 1979).

2. "Biologicals" refers to a category of products that includes vaccines, antigens, antitoxins, toxoids, serums, plasmas, and blood derivatives for human use.

3. Office of Technology Assessment, *A Review of Selected Federal Vaccine and Immunization Policies* (Washington, DC: U.S. Government Printing Office, 1979), 27.

4. Institute of Medicine, *Vaccine Supply and Innovation* (Washington, DC: National Academies Press, 1985), v.

5. Ibid., 53.

6. See Chapter 5 for a more complete discussion of the effects of product liability on industry incentive structures during this time period.

7. Office of Technology, *Review of Selected Federal Vaccine and Immunization Policies*, 150.

8. See Appendix 1. These numbers do not include smallpox vaccine license data; these licenses were cancelled in the late 1970s in response to eradication of the disease, not in response to higher regulatory standards.

9. Margaret Hamburg, "Innovation, Regulation, and the FDA," *New England Journal of Medicine* 363 (December 2, 2010): 2228–2232.

10. Esther Schmid and Dennis Smith, "Is Declining Innovation in the Pharmaceutical Industry a Myth?" *Drug Discovery Today* 10, no. 15 (August 2005): 1034.

11. The FDA Modernization Act of 1997 consolidated establishment and product license requirements into a single Biologic License Application (BLA) to reduce the amount of information that producers had to file with the FDA. This legislation also streamlined the application process for manufacturing changes.

12. Lance Gordon estimated that manufacturers often invest $30 million or more in production facilities before biologics receive FDA approval. Grabowski and Vernon, *The Search for New Vaccines* (Washington, DC: AEI Press, 1997) p. 27.

13. Ibid, 28.

14. R. Gordon Douglas, "The Vaccine Industry," in *Vaccines*, 4th ed., ed. Stanley A. Plotkin and Walter A. Orenstein (Philadelphia: W. B. Saunders, 2003), 47–51. Good Manufacturing Practices (GMP) were first introduced in the 1940s after contaminated sulfathiazole tablets harmed three hundred people. GMP regulations have been continuously updated through the years. Enhanced regulations have made it more difficult for older vaccines, such as Emergent's (formerly BioPort) anthrax vaccine, to be relicensed because these older vaccines had to comply with current GMPs.

15. Michael Kremer and Rachel Glennerster, *Strong Medicine: Creating Incentives for Pharmaceutical Research on Neglected Diseases* (Princeton, NJ: Princeton University Press, 2004); Michael Kremer, "Creating Markets for New Vaccines, Part 1: Rationale" (working paper 7716, National Bureau of Economic Research, Cambridge, MA, May 2000), www.nber.org/papers/w7716; Rachel Glennerster and Michael Kremer, "A Better Way to Spur Medical Research and Development," *Regulation* 23, no. 2 (2002): 34–39.

16. James Sorrentino, former vaccine research scientist at the National Drug Company, interview by author, Swiftwater, PA, May 25, 2001.

17. Grabowski and Vernon, *Search for New Vaccines*, 2.

18. Averages derived from the 2010 CDC price list for pediatric VFC vaccines and adult vaccines, www.cdc.gov/vaccomes/programs/vfc/dcd-vac-price-list.htm (accessed December 15, 2010).

19. Michael Kremer and Christopher Snyder, "Why Are Drugs More Profitable Than Vaccines?" (working paper 9833, National Bureau of Economic Research, Cambridge, MA, July 2003), www.nber.org/papers/w9833.

20. Consolidation is not always bad for innovation. The parent company brings new resources to the acquired vaccine producer, enabling new hires, facility upgrades, and improvements in quality assessment and control. Sharp & Dhome, for example, produced a record number of new vaccines after Merck acquired it in 1953 and hired Maurice Hilleman to run the vaccine division in 1957. After Novartis acquired Chiron in 2005, they licensed two new flu vaccines in 2009 and a conjugate meningococcal vaccine, Menveo, in 2010.

21. Robert Pear, "Juvenile Vaccine Problems Worry Officials and Doctors," *New York Times*, December 2, 2001. Bruce Gellin, director of the National Vaccine Program Office, notes that growing supply problems may also be a result of the fact that a larger number of vaccines are now in routine use. Correspondence with author, December 13, 2010.

22. The federal government reported shortages of diphtheria, tetanus, pertussis, and pneumococcal vaccines and limited deliveries of influenza, chicken pox, measles, mumps, rubella, and influenza vaccines for the 2001–2002 winter season.

23. Editorial, "Averting Vaccine Disaster," *Washington Post*, February 22, 2002.

24. A "new vaccine" is defined as the first safe and effective vaccine licensed to prevent a disease for which no form of active immunity was previously available. An "improved vaccine" is defined by enhancements to the safety, efficacy, or scope (in the case of multivalent and combination vaccines) of a preexisting vaccine.

25. This particular list also includes product introductions from foreign companies.

26. Grabowski and Vernon, *Search for New Vaccines*, 8.

27. Ibid.

28. Ibid.

29. Ibid., 9. Grabowski and Vernon assert that vaccines account for the majority of products represented in the "biologicals" category.

30. Susan Raigrodski, CBER, interview by author, March 2, 2001.

31. Ibid.

32. See Appendix 1 for a complete explanation of data collection sources and methods.

33. These numbers exclude licenses issued to foreign companies and licenses for vaccine ingredients produced in bulk for further manufacturing use.

34. Despite these corrections and additions, the data presented in Figures 2.1 and 2.2 remain incomplete, especially for the period prior to the passage of the Food, Drug, and Cosmetic Act of 1938.

35. If there were multiple introductions of a new vaccine from different companies in the same year, they count as new vaccines in the first year, but not in subsequent years. I did not use priority review as a metric for vaccine innovation because CBER did not begin to prioritize licensures in this way until 2000. William Egan, personal communication, January 5, 2011. (Egan served as deputy and acting director of CBER from 1999 to 2004.)

36. Examples include: Office of Technology Assessment, *Review of Selected Federal Vaccine and Immunization Policies*; Institute of Medicine, *Vaccine Supply and Innovation*; Grabowski and Vernon, *Search for New Vaccines*; Louis Galambos and Jane Eliot Sewell, *Networks of Innovation: Vaccine Development at Merck, Sharp & Dohme, and Mulford, 1895–1995* (Cambridge: Cambridge University Press, 1995).

37. Alfred D. Chandler, Jr., *Scale and Scope: The Dynamics of Industrial Capitalism* (Cambridge, MA: Harvard University Press, 1990); Leonard S. Reich, *The Making of American Industrial Research: Science and Business at GE and Bell, 1876–1926* (Cambridge: Cambridge University Press, 1985); David A. Hounshell and John Kenly Smith, *Science and Corporate Strategy: Du Pont R & D, 1902–1980* (New York: Cambridge University Press, 1988); Emerson W. Pugh, *Building IBM: Shaping an Industry and Its Technology* (Cambridge, MA: MIT Press, 1984); Giovanni Dosi, "Technological Paradigms and Technological Trajectories: A Suggested Interpretation of the Determinants and Directions of Technological Change," *Research Policy* 11 (1982): 147–162; Richard N. Foster, "Timing Technological Transitions," in *Technology and the Modern Corporation: A Strategic Perspective*, ed. Mel Horwitch (New York: Pergamon, 1986).

38. Galambos and Sewell, *Networks of Innovation*; Office of Technology, *Review of Selected Federal Vaccine and Immunization Policies*; Institute of Medicine, *Vaccine Supply and Innovation*; Grabowski and Vernon, *Search for New Vaccines*.

39. Although the federal government provided financial assistance for civilian vaccine purchases as early as 1955 under the Poliomyelitis Immunization Act of 1955, bulk vaccine purchases under federal contract did not begin until 1966. The civilian market for vaccines was still quite small prior to the 1960s.

40. Jeff Widmer, *The Spirit of Swiftwater: 100 Years at the Pocono Labs* (Swiftwater, PA: Connaught Laboratories, 1997), 45.

41. Don Metzgar, interview by author, Swiftwater, PA, June 21, 2001.

42. Ibid. Metzgar joined the National Drug Company in 1966 when it was a division of Richardson-Merrell. Richardson-Merrell donated this division to the Salk Institute, which, in turn, sold it to Connaught in 1978. After a series of mergers, the division became part of Aventis Pasteur in 1999.

43. Institute of Medicine, *Financing Vaccines in the 21st Century: Assuring Access and Availability* (Washington, DC: National Academies Press, 2004), 128; Institute of Medicine, *The Children's Vaccine Initiative: Achieving the Vision* (Washington DC: National Academies Press, 1993).

44. CDC Vaccine Price List, www.cdc.gov/nip/vfc/cdc_vac_price_list.htm (accessed April 1, 2003). See also Institute of Medicine, *Financing Vaccines*, 128; Institute of Medicine, *Vaccine Supply and Innovation*, 60, Tables 4.9 and 4.10.

45. Institute of Medicine, *Children's Vaccine Initiative*.

46. Grabowski and Vernon, *Search for New Vaccines,* 3, Table 1–1, "Private catalogue prices and Federal contract prices per dose for children's vaccines, 1985–1996 (dollars)."

47. Matthew M. Davis et al., "The Expanding Vaccine Development Pipeline, 1995–2008," *Vaccine* 28 (2010): 1353–1356.

48. Sana Siwolop, Big Steps for Vaccine Industry, *New York Times,* July 25, 2001, C17; Amie Bateson and Matthias M. Bekier, "Vaccines Where They're Needed," special issue, *McKinsey Quarterly* (2001): 103.

49. Institute of Medicine, *Financing Vaccines in the 21st Century,* 128; Institute of Medicine, *Calling the Shots* (Washington DC: National Academies Press, 2000).

50. Mercer Management Consulting, *Lessons Learned: New Procurement Strategies for Vaccines: Final Report to the GAVI Board* (2002), quoted in Institute of Medicine, *Financing Vaccines in the 21st Century,* 108.

51. Biotechnology Industry Organization, www.bio.org/investor/signs/200210num.asp.

52. Grabowski and Vernon, *Search for New Vaccines,* 10.

53. The number of vaccine licenses that represent innovative activity was slightly higher in 2000–2009 than in the previous decade, but still low relative to midcentury innovation levels. This jump is especially modest when one considers that research and development investment trends have intensified in the first decade of the twenty-first century. The total number of early-stage vaccine research firms nearly doubled, along with the pipeline of vaccine candidates. Three of the four major vaccine producers in the United States also expanded their vaccine research and development investments three to fourfold (measured in terms of the number of products in development). However, recent studies suggest that biotechnology has not improved research and development productivity and may have even increased the risk and uncertainty of drug research and development by exploding the possible number of targets. Davis et al., "Expanding Vaccine Development Pipeline"; Gary Pisano, "Can Science Be a Business? Lessons from Biotech," *Harvard Business Review,* October 2006, 114–125.

54. Like the 1979 OTA study, I define a new vaccine as the first safe and effective vaccine licensed to prevent a disease for which no form of active immunity was previously available. This list does not include combination vaccines, such as DTP and MMR, and it does not represent the early development of antitoxins.

55. This observation is consistent with research on innovation networks and clusters. Examples include: David Audretsch and Maryann Feldman, "R&D Spillovers and the Geography of Innovation and Production," *American Economic Review* 86, no. 3 (June 1996): 630–640; Luigi Orsenigo, Fabio Pammolli, and Massimo Riccaboni, "Technological Change and Network Dynamics: Lessons from the Pharmaceutical Industry," *Research Policy* 30 (1991): 485–508; W. Powell Walter et al., "Network Dynamics and Field Evolution: The Growth of Interorganizational Collaboration in the Life Sciences," *American Journal of Sociology* 4 (January 2005): 1132–1205.

56. The military has had this effect on other industries as well. Merrit Roe Smith, ed., *Military Enterprise and Technological Change: Perspectives on the American Experience* (Cambridge, MA: MIT Press, 1987), 8–9.

3. Vaccine Development during World War II

1. Allan Chase, *Magic Shots* (New York: William Morrow, 1982), 197.

2. Alfred W. Crosby, *Ecological Imperialism: The Biological Expansion of Europe, 900–1900* (Cambridge: Cambridge University Press, 1986).

3. William H. McNeill, *Plagues and Peoples* (New York: Doubleday, 1976); Hans Zinnser, *Rats, Lice and History* (Boston: Little, Brown, 1934); Ralph H. Major, *Fatal Partners: War and Disease* (New York: Doubleday, Doran, 1941); Paul E. Steiner, *Disease in the Civil War: Natural Biological Warfare in 1861–1865* (Springfield, IL: Charles C. Thomas Pub Ltd., 1968); Alfred W. Crosby, *Epidemic and Peace: 1918* (Westport, CT: Greenwood, 1976).

4. Jeffrey Sartin, "Infectious Diseases during the Civil War: The Triumph of the 'Third Army,'" *Clinical Infectious Diseases* 16 (1993): 580–584; S. B. Hays, foreword to *Preventive Medicine in World War II*, vol. 4, *Communicable Diseases Transmitted Chiefly through Respiratory and Alimentary Tracts*, ed. John Boyd Coates, Jr. (Washington, DC: U.S. Government Printing Office) 1958.

5. N. P. Johnson and J. Mueller, "Updating the Accounts: Global Mortality of the 1918–1920 'Spanish' Influenza Pandemic," *Bulletin of the History of Medicine* 2002; 76:101–115; Thomas Francis, Jr., "Influenza in the U.S. Army Medical Service," in *Preventive Medicine in World War II*, 4:85–87; Department of Defense, *Addressing Emerging Infectious Disease Threats: A Strategic Plan for the Department of Defense* (Washington, DC: Walter Reed Army Institute of Research, 1998), 23.

6. Crosby, *Epidemic and Peace.*

7. Francis, "Influenza," 4:85–87. See also Theodore E. Woodward, *The Armed Forces Epidemiological Board: Its First 50 Years* (Falls Church, VA: U.S. Army Medical Department, 1990).

8. *History of the Relation of the SGO to BW Activities*, NA: RG 112, E. 295A, B. 13.

9. *Digest of Information Regarding Axis Activities in the Field of Bacteriological Warfare* (January 8, 1943), NAS: CBW Files.

10. Gustave J. Dammin and Elliott S. A. Robinson, "Medical Laboratories," in *Preventive Medicine in World War II*, vol. 9, *Special Fields*, ed. Robert Anderson (Washington, DC: U.S. Government Printing Office, 1969), 578.

11. *Digest of Information Regarding Axis Activities in the Field of Bacteriological Warfare* (January 8, 1943).

12. An intelligence digest indicates that in 1942 another Japanese doctor attempted to obtain the virus from a lab in Brazil. *Digest of Information Regarding Axis Activities in the Field of Bacteriological Warfare* (January 8, 1943).

13. Ibid.

14. Reflecting on the inception and administration of the U.S. BW program in World War II, Ernest Goodpasture, a member of the WRS

advisory committee, later concluded, "It seems to me that we have all along acted more by an emotional reaction than by a very critical analysis of the situation." Goodpasture to Perry Pepper, October 16, 1946, NAS: CBW Files.

15. *History of the Relation to the SGO to BW Activities.*

16. Ibid.

17. Dammin and Robinson, "Medical Laboratories," 9:579.

18. Germs As Weapon Feared by Parran, *New York Times*, January 13, 1942.

19. WBC Committee, An Account of Its Initiation and Early Activities, May 15, 1944, NAS: CBW Files.

20. Ibid.

21. WRS development projects were transferred to the Chemical Warfare Service (CWS) in 1943.

22. This board became known as the Army Epidemiology Board in 1946. In 1949, when the board became responsible for the navy and the air force, it was renamed the Armed Forces Epidemiology Board.

23. BICIED oversaw the Commissions on Acute Respiratory Diseases, Air-Borne Infections, Epidemiological Survey, Hemolytic Streptococcal Infections, Influenza, Measles and Mumps, Meningococcal Meningitis, Neurotropic Virus Diseases, Pneumonia, and Tropical Diseases.

24. Brig. General S. Bayne-Jones, deputy chief, Preventive Medicine Service, SGO of the U.S. Army and director of U.S.A. Typhus Commission, NA: RG 156, E. 488, B. 183.

25. The CMR was created by executive order in June of 1941 to supplement the efforts of the National Defense Research Committee. Together these two committees formed the OSRD.

26. Vannevar Bush, *Pieces of the Action* (New York: William Morrow, 1970), 209.

27. George W. Merck, "Peacetime Implications of Biological Warfare," *Merck Report* (July 1946), MA.

28. George W. Merck to F. Jewett, president of the NAS, April 21, 1944, NAS: CBW Files.

29. WBC Committee, Account of Its Initiation and Early Activities, May 15, 1944.

30. Memorandum by WRS to Vannevar Bush, "B.W.," February 27, 1942, NA: RG 227, E. 1, B. 35.

31. Ibid.

32. Alfred Newton Richards, quoted in *Advances in Military Medicine*, ed. E. C. Andrus et al., vol. 1 (Boston: Little, Brown, 1948), p. 1iii.

33. For an account of some early attempts of corporate liberals to improve business-government relations in the sciences during this period, see Robert Kargon and Elizabeth Hodes, "Karl Compton, Isaiah Bowman, and the Politics of Science in the Great Depression," *Isis* 76, no. 3 (September 1985): 300–318.

34. Irvin Stewart, *Organizing Scientific Research for War* (Boston: Little, Brown, 1948), 320.

35. Richards, quoted in *Advances in Military Medicine*.

36. Internal memorandum, "Recommendation for Future Expansion of the Swiftwater Laboratories," February 18, 1948, AP.

37. A few vaccines were not produced through commercial channels. The AMS contributed to typhus vaccine development, the CWS produced botulism antitoxin, and the Rockefeller Institute and the PHS produced yellow fever vaccine.

38. Paul Adams, *Health of the State* (New York: Praeger, 1982), 148.

39. Randolph Major to Vannevar Bush, July 21, 1941, NA: RG 227, E. 165, B. 58.

40. Ibid.

41. Hans Molitor to Alfred Newton Richards, April 1944, NA: RG 227, E. 165, B. 58.

42. George W. Merck, "Activities of the U.S. in the Field of Biological Warfare," October 31, 1945, NA: RG 165, E. 488, B. 182.

43. Internal memorandum, National Drug Company, "Remember Pearl Harbor," December 12, 1941, AP.

44. In contrast, industrial relations within the National Defense Research Committee (NDRC), a division of the OSRD, were often strained. One reason for the discrepancy between government partnerships with the biological and pharmaceutical manufacturers relative to, for example, chemical manufacturers, may be that vaccine makers did not protect intellectual property as aggressively. L. Owens, "The Counterproductive Management of Science in the Second World War: Vannevar Bush and the Office of Scientific Research and Development," *Business History Review* 68 (1994): 4.

45. R. R. Patch, "The 'E' Award: What It Means to Us," *National Bulletin of the National Drug Company* 4, no. 1 (January 1944), AP.

46. Ibid.

47. James Conant, president of Harvard University and chairman of the National Defense Research Committee, letter to the editor of the *New York Times*, August 3, 1945, in response to an unfavorable editorial regarding Bush's proposals in *Science: The Endless Frontier*, copy in NA: RG 227, E. 2, B. 1.

48. Vannevar Bush to Karl Compton, "Comments on the Subject of Organization for Research," October 24, 1946, LC: Vannevar Bush Papers, Container 138.

49. Thomas Francis, Jr., and William S. Tillet, "Cutaneous Reactions in Pneumonia: The Development of Antibodies Following the Intradermal Injection of Type-Specific Polysaccharides," *Journal of Experimental Medicine* 52 (1938): 573; L. Felton, "Studies on Immunizing Substances in Pneumococci: Response in Human Beings to Antigenic Pneumococcus Polysaccharides Type I and II," *Public Health Reports* 45 (1930): 1833; G. Ekwurzel et al., "Studies on Immunizing Substances in Pneumococci: Report on Field Tests to Determine the Prophylactic Value of a Pneumococcus Antigen," *Public Health Reports* 53 (1938): 1877.

50. C. M. MacLeod et al., "Prevention of Pneumococcal Pneumonia by Immunization with Specific Capsular Polysaccharides," *Journal of Experimental Medicine* 82, no. 6 (1945): 445.

51. Michael Heidelberger, "A 'Pure' Organic Chemist's Downward Path," *Annual Review of Biochemistry* 48 (1979): 1–21.

52. Ibid.

53. W. M. Stanley, "The Preparation of and Properties of Influenza Virus Vaccines Concentrated and Purified by Differential Centrifugation," *Journal of Experimental Medicine* 81, no. 2 (1945): 193–218; A. R. Taylor et al., "Concentration and Purification of Influenza Virus for the Preparation of Vaccines," *Journal of Immunology* 50 (1944): 291–316.

54. Members of the Commission on Influenza and Other Epidemic Diseases in the Army, "A Clinical Evaluation of Vaccination against Influenza," *Journal of the American Medical Association* 124 (1944): 982–985.

55. Louis Julianelle, quoted in the WRS report for the ABC Subcommittee, "The Research Program of the War Research Service," May 13, 1944, NAS: CBW Files.

56. The ABC Committee (1942–1944) was a joint National Academy of Sciences/National Research Council advisory committee established at the request of George Merck to advise the WRS on scientific matters pertaining to biological warfare.

57. Military investments in offensive biological research and development programs began to accelerate in December 1943. Members of the OSS were concerned that Germany's new V-1 rockets might contain botulinum toxin within the warheads. In response, the offensive program was transferred from the WRS to the CWS within the War Department as plans were made for the large-scale development of botulinum toxoid. Under this new arrangement, the Camp Detrick program expanded to include field-testing facilities and production plants in addition to research and development laboratories. At the height of the program, approximately 3,900 personnel were associated with the biowarfare defense mission of the special projects division of the CWS. U.S. Department of the Army, *U.S. Army Activity in the U.S. Biological Warfare Programs*, vol. 1, February 24, 1977.

58. Goodpasture to Pepper, October 16, 1946.

59. Eric von Hippel, *Democratizing Innovation* (Cambridge, MA: MIT Press, 2006), 107.

60. Ibid., 109.

61. Ibid.

62. Ibid.

63. Memorandum to commanding general, U.S. Army Medical R&D Command, "Expanded Infectious Disease Research Program," April 17, 1959, NA: RG 112, E. 1004, B. 90.

64. Ibid.

65. Other studies have noted political and professional motivations for this separation of preventive and curative medicine. Evelynn Hammonds's history of diphtheria prevention in New York City at the turn of the century highlights this point. Hammonds, *Childhood's Deadly Scourge: The Campaign to Control Diphtheria in New York City, 1880–1930* (Baltimore: Johns Hopkins University Press, 1999), 207. Paul Starr attributes this separation to an abiding

American belief that "the state should not interfere with private business." Starr, *The Social Transformation of American Medicine* (New York: Basic Books, 1982), 196.

66. John E. Gordon, "General Considerations of Modes of Transmission," in *Preventive Medicine in World War II*, 4:5.

67. Francis, "Influenza," 4:121.

68. Internal memorandum, National Drug Company, "Recommendation for Future Expansion of the Swiftwater Laboratories," February 18, 1948.

69. The SGO was familiar with Heidelberger through some contract research he performed for the WRS during the war. He had participated in the WRS blood studies in which scientists were contracted to test blood samples of war prisoners for evidence of antibodies to anthrax and botulism antitoxin.

70. Michael Heidelberger et al., "The Human Antibody Response to Simultaneous Injection of Six Specific Polysaccharides of Pneumococcus," *Journal of Experimental Medicine* 88 (1948): 369.

71. Office of Technology Assessment, *A Review of Selected Federal Vaccine and Immunization Policies* (Washington, DC: U.S. Government Printing Office, 1979), 32.

72. William H. McNeill, *The Pursuit of Power: Technology, Armed Force, and Society since A.D. 1000* (Chicago: University of Chicago Press, 1982), 360.

73. Atomic weapons research suffered from moral ambiguity as well. Unlike biological weapons programs, however, the Manhattan Project did not contract a large number of physicians that had taken the Hippocratic oath.

74. Rene Dubos spent the majority of his career as a microbiologist at Rockefeller University. From 1942 to 1944, he was a professor of pathology and tropical medicine at Harvard Medical School.

75. They devised a method of forced aeration in which flowing air columns permitted complete oxidation of the nutrients to accelerate culture growth. Dubos's group could obtain maximum yields with this method within twelve hours. Rene Dubos, reporting in Merck, "Activities of the U.S. in the Field of Biological Warfare," October 31, 1945.

76. Karl Meyer to E. B. Fred, December 18, 1942, NAS: CBW Files.

77. WRS report for the ABC Subcommittee, "The Research Program of the War Research Service," May 13, 1944.

78. Ibid.; Historical Report of WRS, November 1944–Final, NA: RG 165, B. 185.

79. Vannevar Bush to Alfred Newton Richards, May 23, 1949, LC: Vannevar Bush Papers, Container 97.

80. Merck, "Peacetime Implications of Biological Warfare," *Merck Report* (July 1946).

81. For an examination of the justifications employed by British scientists engaged in offensive biological research activities, see B. Balmer, "Killing 'without the Distressing Preliminaries': Scientists' Defence of the British Biological Warfare Programme," *Minerva* 40 (2002): 57–75.

4. Wartime Legacies

1. Alfred D. Chandler, Jr., "The Competitive Performance of U.S. Industrial Enterprises since the Second World War," *Business History Review* 68 (Spring 1994): 1–72; D. Mowery and N. Rosenberg, *Paths of Innovation* (Cambridge: Cambridge University Press, 1998).

2. Richard Slee, quoted in Jeff Widmer, *The Spirit of Swiftwater* (Swiftwater, PA: Connaught Laboratories, Inc., 1997).

3. Internal memorandum National Drug Company, "A Recommendation for Future Expansion of the Swiftwater Laboratories," February 18, 1948, AP.

4. Memorandum by the secretary of war to the chief of staff, "Research and Development in Biological Warfare" (September 13, 1945), NA: RG 165, E. 488, B. 182.

5. Ralph Parker to Rolla Dyer, director of NIH (July 15, 1948), NA: RG 443, Decimal File #0470.

6. Although the postwar period represents a time of relative cultural acceptance of vaccines, antivaccinationist groups had been active since the development of the first smallpox vaccine in 1796. For an account of antivaccination movements at the turn of the century, see Judith W. Leavitt, *The Healthiest City: Milwaukee and the Politics of Health Reform* (Princeton, NJ: Princeton University Press, 1982). See also Arthur Allen, *Vaccine: The Controversial Story of Medicine's Greatest Lifesaver* (New York: W. W. Norton, 2007).

7. Internal memorandum, National Drug Company, "A Recommendation for Future Expansion of the Swiftwater Laboratories," February 18, 1948.

8. Ibid.

9. Ibid.

10. Jeff Widmer, *Spirit of Swiftwater*, 44.

11. Bruce L. R. Smith, *American Science Policy since World War II* (Washington, DC: Brookings Institution, 1990).

12. Tom Whayne and Joseph McNich, "Fifty Years of Medical Progress," *New England Journal of Medicine* 244 (1951): 592–601.

13. Draft Report of the OSRD Medical Advisory Committee, "Effect of War on Medical Education and Research," submitted by Walter Palmer, chairman of the committee, to Vannevar Bush, April 25, 1945, NA: RG 227, E. 2, B. 1.

14. Brig. General James Simmons, chief of the Preventive Medicine Service, OSG, U.S. Army, statement for presentation before the Senate Sub-Committee on Wartime Health and Education, December 14, 1944, NA: RG 165, E. 488, B. 183.

15. Ibid.

16. Smith, *American Science Policy since World War II.*

17. For a historical account of the politics surrounding the acceptance of a national science program during this period, see D. J. Kevles, "Principles and Politics in Federal R&D Policy, 1945–1990: An Appreciation of the Bush Report," preface to Vannevar Bush, *Science: The Endless Frontier* (1945; repr., Washington, DC: National Science Foundation, 1990). See also Daniel Lee Kleinman, *Politics on the Endless Frontier* (Durham, NC: Duke University Press, 1995).

18. Bush, *Science: The Endless Frontier.*

19. Draft Report of the OSRD Medical Advisory Committee, "Effect of War on Medical Education and Research," submitted by Walter Palmer, chairman of the committee, to Vannevar Bush, April 25, 1945.

20. Minutes from the Surgeon General's Early Morning Meetings, July 21, 1961, NA: RG 112, E. 1019.

21. For a full list of vaccines and the nature of military contributions, see Appendix 3.

22. Ibid.

23. The AEB was renamed the AFEB in 1949 when it became responsible for navy and air force research activities as well.

24. See Appendix 3.

25. Memo to the director of WRAIR, "Program for Expanded Research on Infectious Disease," July 14, 1959, NA: RG 112, E. 1035, B. 90.

26. John Ellis van Courtland Moon, "The US Biological Weapons Program," in *Deadly Cultures: Biological Weapons since 1945*, ed. Mark Wheelis, Lajos Rózsa, and Malcolm Dando (Cambridge, MA: Harvard University Press, 2006), 12.

27. Minutes, Spring Meeting of the Armed Forces Epidemiology Board, May 18–20, 1959, WR.

28. Minutes from the Surgeon General's Early Morning Meetings, July 1, 1960, NA: RG 112, E. 1014, B. 9.

29. Memo to the commanding general, U.S. Army Medical R&D Command, "Expanded Infectious Disease Research Program," April 17, 1959, NA: RG 112, E. 1004, B. 90.

30. Leonard R. Friedman, "American Medicine as a Military-Political Weapon," *Army R&D Command Annual Report* (1962), NA: RG 112, E. 1012.

31. Ibid.

32. Ibid.

33. Peter Hotez, "Vaccine Diplomacy," *Foreign Policy* (May/June 2001): 68–69.

34. Minutes from the Surgeon General's Early Morning Meetings, July 21, 1961.

35. S. Benison, "International Medical Cooperation: Dr. Albert Sabin, Live Polio Virus Vaccine and the Soviets," *Bulletin of the History of Medicine* 56 (1982): 460–480.

36. www.polioeradication.org/casecount.asp.

37. Donald A. Henderson, *Smallpox: The Death of a Disease: The Inside Story of Eradicating a Worldwide Killer* (Amherst, NY: Prometheus Books, 2009).

38. Army Medical Service Research and Development Program, fiscal years 1952–1969, NA: RG 112, E. 1013.

39. Final Report on Review of Medical and Biological Programs within the Department of Defense, Institute of Defense Analysis, August 1962, B-3.

40. Maurice Hilleman, interview by author, West Point, PA, June 22, 2000.

41. WRAIR, fiscal year 1957, NA: RG 112, E. 1035, B. 42.

42. E. W. Grogan, "Report of Temporary Duty Travel," WRAIR Disposition Form, March 3, 1959, NA: RG 112, E. 1004, B. 85.

43. Vannevar Bush (remarks delivered at the 23rd Annual Scientific Assembly of the Medical Society of the District of Columbia, Washington, DC, October 1, 1952), LC: Vannevar Bush Papers, Container 134.

44. Maurice Hilleman, interview by author, West Point, PA, May 8, 2000.

45. Patrick Kelley, director DOD-GEIS, WRAIR, interview by author, Silver Spring, MD, April 14, 2000.

46. Minutes from the Surgeon General's Early Morning Meetings, May 10, 1957, NA: RG 112, E. 1014, B. 6.

47. Maurice Hilleman, Six decades of vaccine development—a personal history, *Nature Medicine Vaccine Supplement* (May 1998): 507. For an excellent review of Hilleman's life and work, see Paul Offit, *Vaccinated: One Man's Quest to Defeat the World's Deadliest Diseases* (Washington, DC: Smithsonian Books, 2007).

48. There are indications that the industry was already making efforts to improve its reputation for sloppy science before the war. By the 1930s, the efforts of a handful of companies to hire a small staff of in-house researchers were improving their reputation among academic scientists. Swann, *Academic Scientists and the Pharmaceutical Industry* (Baltimore: Johns Hopkins University Press, 1988).

49. As the director of the OSRD, Bush had often emphasized the more fundamental aspects of his research agenda in order to reduce the perception of conflicting interests in OSRD contracts. David A. Mindell, *Between Human and Machine: Feedback, Control, and Computing before Cybernetics* (Baltimore: Johns Hopkins University Press, 2004).

50. Vannevar Bush to Alfred Newton Richards, May 23, 1949, LC: Vannevar Bush Papers, Container 97.

51. Ibid.

52. Vannevar Bush, "Science and Business" (remarks delivered at the 10th Annual Rutgers Business Conference, New Brunswick, NJ, MA.)

53. Louis Galambos and Jane Eliot Sewell, *Networks of Innovation: Vaccine Development at Merck, Sharp & Dohme, and Mulford, 1895–1995* (Cambridge: Cambridge University Press, 1995), 38.

54. Memorandum on Planning Activities of Merck and Company by Vannevar Bush, December 8, 1952, LC: Vannevar Bush Papers, Container 72.

55. Jeff Widmer, *Spirit of Swiftwater*, 45.

56. Arthur D. Little, "Diversification Opportunities," report to Merck & Company, March 25, 1955, LC: Vannevar Bush Papers, Container 75.

57. Ibid.

58. Vannevar Bush to Henry Johnstone, senior vice president and member of the board, April 1, 1955, LC: Vannevar Bush Papers, Container 75.

59. Ibid.

60. P. Stryker, "Tricky Work, Being Board Chairman," *Fortune* (May 1960), 202.

61. Vannevar Bush, "Science and Business."

62. "Merck & Co., Inc.," Harvard Business School Case Study (1960), 13, MA.

63. Antonie Knoppers, vice chairman of the board, COO, and president of Merck & Company (1952–1975), interview by author, New York, NY, July 11, 2000.

64. Sharp & Dohme also provided Merck with sales and distribution channels that permitted them to market their own products directly.

65. J. Enders, T. Weller, and F. Robbins, "Cultivation of the Lansing Strain of Poliomyelitis Virus in Cultures of Various Embryonic Tissues," *Science* 109 (1949): 85–87. Building on the tissue culture techniques developed by Goodpasture in the 1930s, the Enders group developed a method for growing viruses in cell cultures. Previously, researchers relied on infected animal organs and tissues to culture viruses, which made it difficult to grow large quantities for vaccine production.

66. Internal report, Tissue Culture Department, "Permanent Quarters for Poliomyelitis Virus Vaccine," May 12, 1955, LC: Vannevar Bush Papers, Container 75.

67. Maurice Hilleman, "Personal Historical Chronicle of Six Decades of Basic and Applied Research in Virology, Immunology, and Vaccinology," *Immunological Reviews* 170 (1999): 7–27.

68. William H. McNeill, *The Pursuit of Power: Technology, Armed Force, and Society since A.D. 1000* (Chicago: University of Chicago Press, 1982), 373.

69. Vannevar Bush, quoted in G. Pascal Zachary, *Endless Frontier* (New York: Free Press, 1997), 238.

70. Vannevar Bush and J. Conant to H. Stimson, October 27, 1944, quoted in Zachary, *Endless Frontier*, 238.

71. Vannevar Bush, "Of What Use Is a Board of Directors?," 3, MA.

72. Vannevar Bush, "Science and Business." Bush struggled to balance his patriotic support of national objectives with his desire to preserve free markets. While he advocated closer business-government collaboration for national defense, he urged industry to seek collaborative arrangements that did not compromise their long-term strategic goals. To this end, he urged industry to demand an equal voice in the early conception of the project. "The judgment as to the justification for carrying on the program, [and] the estimate of the value of the potential results," he warned, "cannot just be left to some military committee in Washington. Unless the company is admitted initially into confidence and given the opportunity to form its own judgment as to values, unless the technical staff can genuinely collaborate with the military in formulation of the program, it had better leave it alone."

73. "What the Doctor Ordered," *Time Magazine*, August 18, 1952.

74. George W. Merck, "Peacetime Implications of Biological Warfare," *Merck Report* (July 1946), MA.

75. Hilleman, "Personal Historical Chronicle," 12.

76. Galambos and Sewell, *Networks of Innovation*, 62–63.

77. Maurice Hilleman, interview by author, West Point, PA, May 8, 2000.

78. Ibid.

79. Vannevar Bush (remarks delivered at the 23rd Annual Scientific Assembly of the Medical Society of the District of Columbia, Washington, DC, October 1, 1952).

80. Maurice Hilleman, interview by author, West Point, PA, May 8, 2000.

81. Ibid.

82. Maurice Hilleman and J. Werner, "Recovery of New Agents from Patients with Acute Respiratory Illness," *Proceedings of the Society of Experimental Biological Medicine* 85 (1954): 183–188. Similar observations were being made in other locations at this time. Rowe and Huebner detected a transmissible cytopathogenic agent in surgically removed tonsil and adenoidal tissue cultures of children. R. Huebner et al., "Adenoidal-Pharyngeal Conjunctival Agents," *New England Journal of Medicine* 251 (1954): 1077–1087.

83. Maurice Hilleman et al., "Appraisal of Occurrence of Adenovirus Caused Respiratory Illness in Military Populations," *American Journal of Hygiene* 66 (1957): 29–51.

84. Offit, *Vaccinated*, 130.

85. Maurice Hilleman, interview by author, West Point, PA, May 8, 2000.

86. Ibid.

87. T. Campbell, "Reflections on Research and the Future of Medicine," *Science* 153, no. 3734 (July 22, 1966): 446.

88. Maurice Hilleman, "Some Thoughts on Industrial Research in the Health Sciences," (remarks, Industrial Research Institute, Cincinnati, OH, October 20, 1975), MA.

89. Ibid.

90. Maurice Hilleman, interview by author, West Point, PA, May 8, 2000.

91. "Licensure Dates for Products Developed by Virus and Cell Biology Research," 1996, MA.

92. Maurice Hilleman, interview by author, West Point, PA, May 8, 2000.

93. Ibid.

94. Brig. General James Simmons, chief of the Preventative Medicine Service, OSG, U.S. Army, statement for presentation, Before the Senate Sub-Committee on Wartime Health and Education, December 14, 1944, NA: RG 165, E. 488, B. 183.

95. N. Vedros et al., "Studies on Immunity in Meningococcal Meningitis," *Military Medicine* 131, no. 11 (1966): 1413–1417.

96. W. Bell and D. Silber, "Meningococcal Meningitis: Past and Present Concepts," *Military Medicine* 136, no. 7 (1971): 601–611; M. Artenstein et al., "Immunoprophylaxis of Meningococcal Infection," *Military Medicine* 139, no. 2 (1972): 91–95; E. Gotschlich et al., "Human Immunity to the Meningococcus III: Preparation and Immunochemical Properties of the Group A, Group B, and Group C Meningococcal Polysaccharides," *Journal of Experimental Medicine* 129 (1969): 1349–1365.

97. Interview with Maurice Hilleman, West Point, PA, May 8, 2000.

98. Ibid.

99. Sorrentino joined the National Drug Company in 1967 as assistant manager of Viral Products. He was promoted to director of Manufacturing and Development in the early 1970s. In 1974, he joined Richardson-Merrell

(parent company of National Drug) where he eventually became director of research at Richardson-Vicks for over-the-counter respiratory medicine. Sorrentino is currently the managing director of Healthcare Products Development Inc. in Norwalk, CT.

100. James Sorrentino, interview by author, Norwalk, CT, May 25, 2001.

101. Ibid.

102. Don Metzgar joined the National Drug Company as a senior virologist in 1966. He became the director of Biological Research and Manufacturing in the mid-1970s and was appointed vice president of operations in 1978. He retired in 1994 after serving as senior vice president of Connaught Laboratories in 1994. (National Drug's parent company, Richardson-Merrell, donated the facilities of the former National Drug Company to the Salk Institute in 1978. Connaught purchased these facilities from the Salk Institute in that same year.)

103. James Sorrentino, interview by author, Norwalk, CT, May 25, 2001.

104. Ibid.

105. Ibid.

106. According to Metzgar, Lilly abandoned the project because, unlike National, Lilly did not have the automated egg-handling equipment that Sorrentino had recently developed to supply the high volume of material required for zonal centrifugation.

107. James Sorrentino, interview by author, Norwalk, CT, May 25, 2001.

108. Ibid.

109. Ibid.

110. Ibid.

111. Ibid.

112. Don Metzgar, interview by author, Swiftwater, PA, June 21, 2001.

113. James Sorrentino, interview by author, Norwalk, CT, May 25, 2001.

114. Ibid.

115. Ibid.

116. Don Metzgar, interview by author, Swiftwater, PA, June 21, 2001.

117. Ibid.

118. Ibid.

119. Sorrentino thought that Merck and National shared more than efficacy data under this arrangement. He believed that "there was a problem with polysaccharide size." He explained, "What I've been told is that we had purified polysaccharide that was appropriately sized and Merck was having trouble so that data was shared so everybody could produce the appropriate amount of polysaccharide with the appropriate size." James Sorrentino, interview by author, Norwalk, CT, May 25, 2001.

120. Ibid.

121. Don Metzgar, interview by author, Swiftwater, PA, June 21, 2001.

122. Martha L. Lepow, Irving Goldschneider, Ronald Gold, Martin Randolph, and Emil C. Gotschlich, "Persistence of Antibody Following Immunization of Children With Groups A and C Meningococcal Polysaccharide Vaccines," *Pediatrics* 60, no. 5 (1977): 673–680.

123. Merck discontinued production of the meningitis vaccine in 1995.

124. Metzgar attributes this trend to overcrowded dormitories: "Colleges and universities are different now. When I went to Purdue, there were twelve thousand students on the whole campus. Now there are sixty-four thousand, and they are cramped in small spaces and it looks more like the military." Don Metzgar, interview by author, Swiftwater, PA, June 21, 2001.

125. M. Mauss, *The Gift: Forms and Functions of Exchange in Archaic Societies* (New York: Free Press, 1954). Sociologists refer to the same distinction as social versus market exchanges. See Russell Hardin, *Collective Action* (Baltimore: Johns Hopkins University Press, 1982), 206.

126. For a review of the literature of economic anthropology as applied to modern scientific exchange relationships, see Warwick Anderson, "The Possession of Kuru: Medical Science and Biocolonial Exchange," *Society for Comparative Study of Society and History* 10 (2000): 713–744.

127. The free exchange observed in gift economies is also characteristic of "communities of practice" described in organizational studies: J. S. Brown and P. Dugid, "Organizational Learning and Communities-of-Practice: Toward a Unified View of Working, Learning, and Innovation," *Organization Science* 2, no. 1 (1991): 40–57.

128. Smith, *American Science Policy since World War II*, 72.

129. Galambos and Sewell, *Networks of Innovation*, 148.

130. James Sorrentino, interview by author, Norwalk, CT, May 25, 2001.

5. The End of an Era

1. Office of Technology Assessment, *Review of Selected Federal Vaccine and Immunization Policies*; Institute of Medicine, *Vaccine Supply and Innovation* (Washington, DC: National Academies Press, 1985).

2. For an account of the circumstances and personalities surrounding this decision, see S. Weiner, *Swine Flu Fever in America: CDC Decides* (working paper), MIT Center for International Studies (MIT, Cambridge, MA, 2007).

3. For an account of these deliberations see Richard E. Neustadt and Harvey V. Fineberg, *The Epidemic That Never Was: Policy-Making and the Swine Flu Affair* (Toronto: Vintage, 1983).

4. "Vaccine Plan: Produce Now, Dicker Later," *Chemical Week* (April 14, 1976).

5. Ibid., 70.

6. "Snag in Flu Vaccine," *Business Week*, June 28, 1976; "Troubles Plague Flu Vaccine," *Business Week*, June 7, 1976.

7. Neustadt and Fineberg, *The Epidemic That Never Was*, 89.

8. Ibid., 86.

9. David F. Gordon, Don Noah, and George Fidas, "The Global Infectious Disease Threat and Its Implications for the United States," *National Intelligence Estimate* 99-17D (January 2000): Figure 17: Trends in Infectious Disease Mortality Rates in the United States.

10. Don Metzgar, interview by author, Swiftwater, PA, June 21, 2001.

11. Edward W. Kitch, "Vaccines and Product Liability: A Case of Contagious Litigation," *Regulation* (May/June 1985), 13.

12. Louis Galambos and Jane Eliot Sewell, *Networks of Innovation: Vaccine Development at Merck, Sharp & Dohme, and Mulford, 1895–1995* (Cambridge: Cambridge University Press, 1995).

13. Internal memorandum by David William, VP and general manager, Connaught Laboratories, Swiftwater, PA, 1985, AP.

14. Ibid.

15. Jeff Widmer, *The Spirit of Swiftwater: 100 Years at the Pocono Labs* (Swiftwater, PA: Connaught Laboratories, 1997), 61.

16. The Canadian firm Connaught eventually inherited National Drug's laboratories in Swiftwater, PA, after Merrell-National donated the company to the Salk Institute in 1978.

17. Neustadt and Fineberg, *The Epidemic That Never Was*, 71.

18. "A Risky Exodus from Vaccines," *BusinessWeek*, April 10, 1978, 118.

19. Ibid.

20. Robert Hendrickson, oral history, interview by Jeffrey Sturchio and Louis Galambos, December 20, 1991, and April 13, 1992, MA.

21. Ibid.

22. Consolidation continued over the years. Wyeth and Lederle merged as a division under American Home Products, and French manufacturer Aventis-Pasteur absorbed Connaught.

23. J. Clark and R. Ghislain, "Commercial Aspects of the Vaccine Industry," in *New Generation Vaccines*, ed. G. Woodrow and M. Levine (New York: Marcel Dekker, 1990).

24. Don Metzgar, interview by author, Swiftwater, PA, June 21, 2001.

25. Ibid.

26. This program put a surcharge on vaccine sales to create a federal fund from which individuals could be compensated.

27. Ronald A. Rader, *Biopharma: Biopharmaceutical Products in the U.S. Market* (Rockville, MD: Biotechnology Information Institute, 2001).

28. R. Arnould and L. DeBrock, "The Application of Economic Theory to the Vaccine Market," in *Supplying Vaccine: An Economic Analysis of Critical Issues*, ed. M. Pauly (Amsterdam: IOS Press, 1996), 103.

29. Lance Gordon, prepared statement before the House Committee on Commerce, June 15, 1995.

30. M. Sing and M. Willian, background paper (presented at A Study of the Economic Underpinnings of Vaccine Supply, sponsored by the National Vaccine Program Office and the CDC, Washington, DC, November 12, 1993).

31. P. Roy Vagelos and Louis Galambos, *The Moral Corporation* (Cambridge: Cambridge University Press, 2006), 127.

32. Ibid., 128.

33. Douglas MacMaster, president of Merck, Sharp & Dohme Division, summary testimony, hearing on funding of the National Vaccine Injury Compensation Program, Before the Subcommittee on Select Revenue Measures, Ways and Means Committee, U.S. House of Representatives, March 5, 1986.

34. Ibid.

35. Don Metzgar, interview by author, Swiftwater, PA, June 21, 2001.

36. Ibid.

37. Ibid.

38. Ibid. Sorrentino echoed Metzgar's remarks, stating, "It is my understanding that no one collaborates anymore." James Sorrentino, interview by author, Norwalk, CT, May 21, 2001.

39. Theodore E. Woodward, *The Armed Forces Epidemiology Board: Its First 50 Years* (Falls Church, VA: Office of the Surgeon General, Department of the Army, 1990).

40. Memorandum to commanding general, U.S. Army Medical R&D Command, "Expanded Infectious Disease Research Program," April 17, 1959, NA: RG 112, E. 1004, B. 90.

41. Memorandum to the director of WRAIR, "Program for Expanded Research on Infectious Disease," July 14, 1959, NA: RG 112, E. 1004, Box 90.

42. Philip Russell, interview by author, February 23, 2009.

43. The Convention on the Prohibition of the Development, Production, and Stockpiling of Bacteriological (Biological) and Toxin Weapons and Their Destruction, signed 10 April 1972, entered into force 26 March 1975.

44. Jonathan B. Tucker, "A Farewell to Germs: The U.S. Renunciation of Biological and Toxin Warfare, 1969–1970," *International Security* 27, no. 1 (2002): 112.

45. William Creasy, congressional testimony, recorded in SIPRI, *The Prevention of CBW*, vol. 5, *The Problem of Chemical and Biological Warfare* (Stockholm: Almquvist & Wiksell, 1971), 278.

46. According to Jonathan Tucker's study, the biological research and development budget was cut from $20 million per year to $10 million per year after the United States terminated its offensive biological program. Tucker, "Farewell to Germs," 140.

47. Tom Whayne and Joseph McNich, "Fifty Years of Medical Progress," *New England Journal of Medicine* 244 (1951): 592–601.

48. Daniel J. Kevles, "Principles and Politics in Federal R&D Policy, 1945–1990: An Appreciation of the Bush Report," preface to Vannevar Bush, *Science: The Endless Frontier* (1945; repr., Washington, DC: National Science Foundation, 1990).

49. Ibid.

50. Minutes, Meeting of the Armed Forces Epidemiology Board, October 24, 1969, WR.

51. Ibid.

52. Woodward, *Armed Forces Epidemiology Board*, 124.

53. Philip Russell, correspondence with author, April 15, 2009.

54. Dr. Wisseman, Commission on Rickettsial Diseases, Minutes of the Annual Spring Meeting of the Armed Forces Epidemiology Board, 1972, WR.

55. Ibid.

56. F. Ingelfinger, visit to WRAIR, July 30, 1974, WR.

57. Institute of Defense Analysis, Final Report on Review of Medical and Biological Programs within the Department of Defense, August 1962, 69, NA: RG 319, E. 181, B. 1.

58. Ibid.

59. Ibid., L-1.

60. Vannevar Bush, *Pieces of the Action* (New York: William Morrow, 1970).

61. Vannevar Bush, *Modern Arms and Free Men* (Cambridge, MA: MIT Press, 1968).

62. Philip Russell, interview by author, February 23, 2009.

63. Ibid.

64. Philip Russell, correspondence with author, April 15, 2009.

65. Institute of Defense Analysis, Final Report on Review of Medical and Biological Programs within the Department of Defense, August 1962, L-1, NA: RG 319, E. 181, B. 1.

66. Thomas Monath, interview by author, Cambridge, MA, September 18, 2000.

67. Maurice Hilleman et al., "Appraisal of Occurrence of Adenovirus Caused Respiratory Illness in Military Populations," *American Journal of Hygiene* 66 (1957), 29–51.

68. Maurice Hilleman and J. H. Werner, "Recovery of New Agent from Patients with Acute Respiratory Illness," *Proceedings of the Society of Experimental Biological Medicine* 85: 183–188 (1954); R. J. Huebner, W. P. Rowe, and T. J. Ward, "Adenoidal Pharyngeal Conjunctival Agents," *New England Journal of Medicine* 25 (1954): 1077–1087.

69. Maurice Hilleman et al., "Antibody Response in Volunteers to Adenovirus Vaccine and Correlation of Antibody with Immunity," *Journal of Immunology* 80 (1958): 299–307.

70. Philip Russell, interview by author, February 23, 2009. CBER does not have a record of Lederle's adenovirus vaccine license.

71. The virus transmits best among stressed populations and not as well among populations that are merely crowded, such as college dorms.

72. Colonel Arthur P. Long, chief, Preventive Medicine Division, to Thomas Francis, president, AFEB, April 16, 1959, NA: RG 112, E. 1004, B. 93.

73. Ibid.

74. Ibid.

75. Ibid.

76. Ibid.

77. Minutes from the Surgeon General's Early Morning Meetings, April 22, 1960, NA: RG 112, E. 1014, B. 8.

78. Ibid.

79. Lt. Gen. Heaton, surgeon general, to chief of staff, U.S. Army, February 2, 1966, NA: RG 112, E. 1014, B. 7.

80. Alan Cross and Phil Russell, "Diseases of Military Importance," in *Vaccines: A Biography*, ed. Andrew Artenstein (New York: Springer, 2009).

81. Philip Russell, correspondence with author, April 15, 2009.

82. P. Collis et al., "Adenovirus Vaccines in Military Recruit Populations: A Cost-Benefit Analysis," *Journal of Infectious Diseases* 128 (1973): 745–752.

83. Ibid.

84. Minutes, Meeting of the Armed Forces Epidemiology Board, 1972–1973, WR.

85. Floyd Denny, comments, Minutes of the Spring Meeting of the Armed Forces Epidemiology Board, 1972–1973.

86. Minutes, Meeting of the Armed Forces Epidemiology Board, 1972–1973.

87. Thomas Monath, chief scientific advisor, Oravax (later renamed Acambis, now part of Sanofi Pasteur Biologics Co.), interview by author, Cambridge, MA, September 18, 2000.

88. Ibid.

89. Ibid.

90. R. Cutting, comments, Minutes of the Spring Meeting of the Armed Forces Epidemiology Board, 1972–1973.

91. Minutes, Meeting of the Armed Forces Epidemiology Board, 1972–1973.

92. Floyd Denny, comments, Minutes of the Spring Meeting of the Armed Forces Epidemiology Board, 1972–1973.

93. When Richardson-Merrell (National merged with Vick Chemical to become Richardson-Merrell Inc.) divested their biological operations in 1977, they offered to sell this production facility to the army, but the army declined. The Salk Institute operated as a nonprofit on the Swiftwater campus, and Connaught acquired and operated the remaining commercial facilities until Connaught was acquired by Sanofi Pasteur.

94. Cutting, comments, Minutes of the Annual Spring Meeting of the Armed Forces Epidemiology Board, 1972–1973.

95. Philip Russell, correspondence with author, April 15, 2009.

96. Aaron Friedberg, *In the Shadow of the Garrison State: America's Anti-Statism and Its Cold War Grand Strategy* (Princeton, NJ: Princeton University Press, 2000), 293.

97. Philip Russell, "Vaccines for the Protection of the U.S. Forces: Research Success and Policy Failures" (presented to the Institute of Medicine, Committee on a Strategy for Minimizing the Impact of Naturally Occurring Infectious Diseases of Military Importance: Vaccine Issues in the U.S. Military, Washington, DC, March 22, 2000).

98. Patrick Kelley, correspondence with author, December 30, 2010.

99. Joel Gaydos, correspondence with author, January 12, 2011.

100. Philip Russell, interview by author, February 23, 2009.

101. Ibid.

102. Department of Defense, *DOD Vaccine Production Facility Task Force Final Report* (Washington, DC: U.S. Department of Defense, 1991, draft; 1993).

103. Under an IND, the FDA permits companies to begin testing a drug or vaccine on human populations with written consent from the subjects.

104. Department of Defense, *Report on Biological Warfare Defense Vaccine Research and Development Programs* (July 2001), 3.

105. Ibid., 3.

106. Wayne Pisano, "Strengthening the Supply of Routinely Recommended Vaccines in the U.S.: Industry Perspective" (presented at the National Vaccine Advisory Committee Meeting, February 11–12, 2002).

107. Office of the DDR&E DOD In-House RDT&E Activities Report (FY69 and FY00); D. J. DeYoung, "Breaking the Yardstick: The Dangers of Market-Based Governance," *Defense Horizons*, Center for Technology and National Security Policy, National Defense University (May 2009), www.ndu.edu/ctnsp/publications.html. December 2010.

108. Philip Russell, interview by author, June 30, 2010.

109. R. Jones, "Dynamic DynPort," *Frederick News Post Online*, May 22, 2009, www.fredericknewspost.com. December 2010.

110. Institute for Foreign Policy Analysis, *Missile Defense, the Space Relationship, and the Twenty-First Century* (2006), 102. The literature on the management of research and development refers to this ability as the firms' (or labs') absorptive capacity. Researchers note that firms often invest in in-house research and development because they cannot effectively learn or collaborate with outside sources if they lack a working knowledge of the subject themselves. This concept highlights the limits of outsourcing. Wesley M. Cohen and Daniel A. Levinthal, "Absorptive Capacity: A New Perspective on Learning and Innovation," *Administrative Science Quarterly* 35, no. 1 (1990): 128–152.

111. Thomas Monath, chief scientific advisor, Oravax (now Acambis), interview by author, Cambridge, MA, September 18, 2000.

112. National Science Foundation, *Science and Engineering Indicators, 2000*, Appendix Table 6–8, "Federal Obligation for Academic R & D, By Agency: 1970–99," A-393.

113. U.S. Department of Health and Human Services, National Institutes of Health, National Institute of Allergy and Infectious Diseases, The Jordan Report: *Accelerated Development of New Vaccines 1986 Progress Report.*

114. Ibid.

115. Ibid.

116. C. MacLeod, address (presented at the 30th Anniversary Meeting of the Armed Forces Epidemiology Board, Washington, DC, February 18, 1971), WR.

117. Maurice Hilleman, interview by author, West Point, PA, May 8, 2000.

118. Jon Cohen, *Shots in the Dark* (New York: W. W. Norton, 2001), 320.

119. Maurice Hilleman, "The Frustrating Journey toward an AIDS Vaccine" (unpublished draft), MA.

120. Vannevar Bush, president of the Carnegie Institution of Washington, remarks (presented at the 23rd Annual Scientific Assembly of the Medical Society of the District of Columbia, Washington, DC, October 1, 1952), MA.

121. Maurice Hilleman, "The Business of Science and Science of Business in the Quest of an AIDS Vaccine," *Vaccine* 17 (1999): 1211–1222.

122. The Levine Committee represented an interinstitutional "working group" or advisory committee assigned by the NIH Office of AIDS Research in 1995 to review the organization and function of the NIH's AIDS vaccine research projects. Cohen, *Shots in the Dark*, 282.

123. Cohen, *Shots in the Dark*, 282.

124. Ibid., 283.

125. This report lent support to the recommendations of the Presidential Advisory Council on HIV/AIDS (PACHA), which, in turn, led to Clinton's 1997 call for a reorganization of the AIDS vaccine research effort and the establishment of a new AIDS Vaccine Research Center at NIH. President Clinton, commencement address (presented at Morgan State University Balitimore, MD; The White House, Office of the Press Secretary, May 19, 1997).

126. This phenomenon is known as the "tragedy of the anti-commons." M. A. Heller and R. S. Eisenberg, "Can Patents Deter Innovation? The Anticommons in Biomedical Research," *Science* 280, no. 5364 (1998): 698.

127. R. Carlson, *Biology Is Technology: The Promise, Peril, and New Business of Engineering Life* (Cambridge, MA: Harvard University Press, 2010), 186.

128. Ashish Arora and Alfonso Gambardella, "Complementarity and External Linkage: The Strategies of Large Firms in Biotechnology," *Journal of Industrial Economics* 38, no. 4 (June 1990): 361; Gary P. Pisano, "The Governance of Innovation: Vertical Integration and Collaborative Arrangements in the Biotechnology Industry," *Research Policy* 20, no. 3 (June 1991): 237.

129. Luigi Orsenigo, Fabio Pammoli, and Massimo Riccaboni, "Technological Change and Network Dynamics: Lessons from the Pharmaceutical Industry," *Research Policy* 30 (2001): 485–508; Stephen Kobrin, "The Architecture of Globalization: State Sovereignty in a Networked Global Economy," in *Governments, Globalization, and International Business*, ed. John H. Dunning (Oxford: Oxford University Press, 1997), 149–150; Lynn Krieger Mytelka, "Crisis, Technological Change, and the Strategic Alliance," in *Strategic Partnerships: States, Firms, and International Competition* (Rutherford, NJ: Fairleigh Dickinson University Press, 1991), 16–21; DiMasi, Hansen, and Grabowski, *Price of Innovation.*

130. Geoffrey Jones, *The Evolution of International Business: An Introduction* (London: Routledge, 1996), 142.

131. François Chesnais, preface to *Strategic Partnerships*, x.

132. Orsenigo, Pammoli, and Riccaboni, "Technological Change and Network Dynamics," 485–508.

133. Ibid., 485–486.

134. Galambos and Sewell, *Networks of Innovation*, 196–205.

135. Overall vaccine-related investments in biotechnology were likely higher because these figures do not include biotechnology firms that were not members of the Pharmaceutical Research and Manufacturers Association. Office of Technology Assessment, *Commercial Biotechnology: An International Assessment* (Washington, DC: U.S. Government Printing Office, 1984).

136. Maurice Hilleman, interview by author, West Point, PA, May 8, 2000.

137. He conceded that "two to three vaccines were finished after this period but that was just due to a time-lag in the engineering and scale-up." Maurice Hilleman, interview by author, West Point, PA, May 8, 2000.

138. Galambos and Sewell, *Networks of Innovation*, 241.

6. Biodefense in the Twenty-First Century

1. The Bush administration requested $45 billion to fight terrorism and $5.9 billion to fight bioterrorism in fiscal year 2003. This request reflects an estimated 230 percent and 420 percent increase, respectively, over pre-9/11 spending levels. Office of Management and Budget, *Annual Report to Congress on Combating Terrorism* (Washington, DC: U.S. Government Printing Office, 2003).

2. "President Bush's State of the Union Address to Congress and the Nation," transcript reprinted in *The New York Times*, January 30, 2002.

3. The FDA also reissued a license for an anthrax vaccine (BioThrax) to Emergent BioSolutions. The BioThrax license is based on a 1970's version of the vaccine and does not represent innovative activity. Bavarian Nordic developed a safer smallpox vaccine (Imvamune) under BARDA contracts, but this vaccine has not yet been licensed.

4. Soldiers and civilians shared this perspective. Soldiers doubted they would face anthrax in the field and several risked their military careers by refusing the vaccine. Health-care workers also doubted they would encounter smallpox and the smallpox vaccination program never acheived its immunization goals.

5. For a review of the DOD's struggle to engage industry, see Institute of Medicine, *Protecting Our Forces: Improving Vaccine Acquisition and Availability in the U.S. Military* (Washington, DC: National Academies Press, 2002).

6. Ibid.

7. Ibid.

8. "Pharma Industry Promises to Support Government against Bioterrorism Threat," *Chemical Market Reporter* 260, no. 17 (November 5, 2001): 10.

9. J. Gillis, "Scientists Race for Vaccines: Drug Companies Called Key to Bioterror Fight," *Washington Post*, November 8, 2001.

10. Ibid.

11. "Drug Firms Consider Smallpox Vaccine," *Associated Press* online, November 1, 2001.

12. Leslie Wayne and Melody Petersen, "A Muscular Lobby Tries to Shape Nation's Bioterror Plan," *New York Times*, November 4, 2001.

13. Gillis, "Scientists Race for Vaccines."

14. "Vaccines May Be Profit Engine for Baxter," *Reuters*, November 29, 2001.

15. J. Gillis, "Drugmakers Step Forward in Bioterror Fight," *Washington Post*, October 31, 2001.

16. Carl Feldbaum, "National Security Is at Stake; Biodefense Products Remain Stalled," *Washington Times*, September 10, 2003, A 21.

17. Testimonies before the House Government Reform Committee, Veterans Affairs, and International Relations Subcommittee, *Bioterrorism Vaccines*, October 23, 2001.

18. National Institutes of Health Appropriations testimony for FY 2004, www.asmusa.org/pasrc/nih/2004.html.

19. Una Ryan, testimony before the Subcommittee on Science, Technology, and Space, Senate Committee on Commerce, Science, and Transportation, February 2, 2002.

20. It was later determined that the anthrax sent through the mail was sensitive to an array of antibiotics already available under generic labels.

21. These regulations criminalize the unauthorized possession, use, or transfer of forty-nine select pathogens listed by the Department of Health and Human Services.

22. John Dudley Miller, "Butler's Last Stand," *Scientist* 18, no. 44 (March 1, 2004).

23. Kendall Hoyt and Stephen Brooks, "A Double-Edged Sword: Globalization and Biodefense," *International Security* 28, no. 3 (Winter 2003/2004): 123–148.

24. Rick Smith, interview by author, Swiftwater, PA, March 6, 2003.

25. Quoted in Scott Shane, "Terror Threat Casts Chill over World of Biological Research," *Baltimore Sun*, January 26, 2003.

26. Quoted in David Ruppe, "New Regulations on Biological Materials Get Mixed Reviews," *Global Security Newswire*, December 12, 2002, www.nti.org/d_newswire/issues/newswires/2002_12_12.html (accessed April 2003).

27. Walter Vandersmissen, Government Affairs Director, SmithKline-Beecham (presented at the Albert B. Sabin Vaccine Institute Colloquium, Cold Spring Harbor, NY, December 5–7, 1999), in William Muraskin, ed., *Vaccines for Developing Economies: Who Will Pay?* (New Canaan, CT: The Albert B. Sabin Vaccine Institute, 2001), 58–59.

28. Rep. Christopher Shays, CT, "Defense Vaccines: Force Protection or False Security?" Hearing before the U.S. House, Committee on Government Reform, 106th Cong., 1st Sess., October 12, 1999.

29. Ibid.

30. Charles Schumer, press release, October 16, 2001, www.senate.gov/~schumer/

31. Bernard Sanders, Before House Government Reform Committee, Veterans Affairs and International Relations Subcommittee, *Bioterrorism Vaccines*, October 23, 2001.

32. Judith Miller, "Bush to Request a Major Increase in Bioterror Funds," *New York Times*, February 4, 2002.

33. Ibid.

34. The NIH's Vaccine Research Center (VRC) provides a template for a parallel top-down governance structure within the NIH. The VRC grew out of the 1996 Levine Report to the NIH Office of AIDS Research, which advocated a targeted approach to the development of the AIDS vaccine to bridge the gap between basic and applied vaccine research initiatives. While the VRC is part of the NIH campus, and its plant and equipment are government owned, it is operated by a private contractor (SAIC-Fredrick, Inc). The original aim of the VRC was to drive fundamental research on the AIDS vaccine through late-stage development and into clinical trials. The center coordinates approximately fifty scientists already working on problems that relate to the de-

velopment of an HIV/AIDS vaccine. Much like many midcentury vaccine development programs, the VRC is designed to integrate the full range of research and development activities, including immunology, virology, molecular biology, vaccine design, clinical trials, and production. The VRC concept has expanded to include work on Ebola, Marburg, and influenza vaccines.

35. This plan called for: (1) A next-generation smallpox vaccine that incorporates the modified vaccinia Ankara (MVA) strain. This vaccine could be safely administered to higher-risk populations with compromised immune systems or skin conditions. (2) An anthrax vaccine that incorporates a recombinant protective antigen (rPA). This vaccine would require fewer doses per person and may serve as an effective postexposure prophylaxis. (3) The production of more botulinum antitoxin. (4) A botulism vaccine. (5) A monoclonal antibody botulism therapeutic. (6) A plague vaccine. (7) An ebola vaccine. "President Details Project BioShield," White House press release, February 3, 2003, www.whitehouse.gov/news/releases/2003/02/20030203.html.

36. Marilyn Chase, "Project BioShield Is a Big Incentive to Vaccine Makers," *Wall Street Journal*, January 30, 2002, D2.

37. Chuck Ludlam, author of BioShield II and former senior counsel to Senator Joe Lieberman, reported: "A high-ranking Administration official admitted that it proposed BioShield I solely to protect its right flank when Senator Lieberman was running for President, not as part of a serious bioterror strategy. It's obvious that BioShield I was poorly calculated and the industry response to it has been to yawn." Chuck Ludlam, answers to subcommittee questions roundtable, "When Terror Strikes—Preparing an Effective and Immediate Public Health Response," Subcommittee on Bioterrorism and Public Health Preparedness, July 14, 2005.

38. Congressional Budget Office Cost Estimate, S. 15, Project BioShield Act of 2003, May 7, 2003.

39. Matheny, Mair, and Smith, "Cost/Success Projections." Subsequent studies have used more conservative measures, estimating the actual cost to be between $6.3 billion and $11.6 billion for 2009–2015. L. Klotz and A. Pearson, "BARDA's Budget," *Nature Biotechnology* 27, no. 8 (August 2009): 698.

40. Author correspondence with Chuck Ludlam, June, 23, 2003.

41. N. Munro, "Bioterrorism Plan Expected This Summer," *National Journal*, May 2, 2002.

42. Una Ryan, testimony, Before the House Government Reform Committee, Veterans Affairs and International Relations Subcommittee, *Bioterrorism Vaccines*, October 23, 2001.

43. A. Gambardella, *Science and Innovation: The U.S. Pharmaceutical Industry in the 1980s* (Cambridge: Cambridge University Press, 1995), 147.

44. Acambis became the Sanofi Pasteur Biologics Company in 2008.

45. For an account of the DOD's struggle with BioPort, see Judith Miller, Stephen Engelberg, and William Broad, *Germs: Biological Weapons and America's Secret War* (New York: Simon and Schuster, 2002). BioPort became Emergent BioSolutions in 2004.

46. Jill Wechsler, "Vaccine Shortages Put Spotlight on Production," *Biopharm* 15, no. 5 (2002): 44.

47. Scott Lilly, "Getting Rich on Uncle Sucker: Should the Federal Government Strengthen Efforts to Fight Profiteering?" Center for American Progress, October 20, 2010, www.americanprogress.org/issues/2010/10/uncle_sucker.html.

48. Ibid.

49. VaxGen's CEO later acknowledged that he estimated his company had a 10 percent chance of being able to fill their BioShield contract. K. Rhodes, *Project BioShield: Actions Needed to Avoid Repeating Past Mistakes* (Washington, DC: Government Accountability Office, October 23, 2007).

50. T. Monath, "Industry Involvement in Federal Vaccine Development and Procurement Efforts" (presentation at the Second Meeting of the Institute of Medicine Committee on a Strategy for Minimizing the Impact of Naturally Occurring Diseases of Military Importance: Vaccine Issues in the U.S. Military, 2000).

51. Ibid.

52. I. Kola and J. Landis, "Can the Pharmaceutical Industry Reduce Attrition Rates?" *Nature Reviews Drug Discovery* 3 (2004): 711–715.

53. Increasingly, integrated research and development practices can flourish in large research, development, and manufacturing companies like Merck and Aventis-Pasteur. In recent years, the biopharmaceutical industry has begun to reverse this trend toward flat organizations. According to an independent report on biological warfare defense vaccine R&D programs, "Companies having the capacity and capability tried outsourcing manufacturing but have since pulled these operations back in-house. Unlike outsourced manufacturing of chemical pharmaceuticals, outsourcing of vaccine manufacturing was found to be fraught with difficulties, inordinate process control risks, and added overall costs. As a result, the major vaccine producers limit or do not outsource manufacturing at all." The report notes that "although there are validated technological processes for controlling the manufacturing process for a vaccine, repetition of the process and an element of art in the underlying S&T seem crucial to success." (Department of Defense, *Report on Biological Warfare Defense Vaccine Research and Development Programs* [July 2001], 7.)

54. Frank Gottron, "Project BioShield: Authorities, Appropriations, Acquisitions, and Issues for Congress," *Congressional Research Service*, January 22, 2010, 9.

55. J. Tucker, "Developing Medical Countermeasures: From BioShield to BARDA," *Drug Development Research* 70 (2009): 231.

56. PHEMCE Implementation Plan, *Federal Register* 72, no. 77 (April 23, 2007): 20117–20128.

57. Bob Graham and Jim Talent, *World at Risk: The Report of the Commission on the Prevention of WMD Proliferation and Terrorism* (New York: Vintage, 2008). Senator Bob Graham and Senator Jim Talent, chairman and vice chairman of the Commission on the Prevention of Weapons of Mass Destruction Proliferation and Terrorism, to President Barack Obama, June 7, 2009, published in "Graham and Talent Call on President Obama to Appoint Vice President to Lead Campaign on WMD Proliferation and Terrorism," *Business Wire*, June 8, 2009.

58. Ibid. They argue for putting Vice President Joe Biden in charge of biodefense and prevention. Alternatively, or additionally, the administration could create a senior position within the National Security Council to coordinate crosscutting presidential initiatives for biodefense and prevention.

59. Sidney Altman et al., "An Open Letter to Elias Zerhouni," *Science* 307: 1409 (March 4, 2005).

60. Ceci Connolly, "Bush Promotes Plans to Fight Bioterrorism," *Washington Post*, February 6, 2002, A03. (White House request released February 4, 2002.)

61. Vannevar Bush, speaking before the Select Committee on Government Research of the U.S. House of Representatives, "Vannevar Bush Speaks," *Science* 142, no. 3600 (December 27, 1963), 1623. He made a related observation in his book several years later: "With the Federal government plunging into the support of research on an enormous scale there is danger of the encouragement of mediocrity and grandiose projects, discouragement of individual genius, and hardening of administrative consciences in the universities." Vannevar Bush, *Modern Arms and Free Men* (Cambridge, MA: MIT Press, 1968), 247.

62. Vannevar Bush, *Pieces of the Action* (New York: William Morrow, 1970), 68.

7. *The Search for Sustainable Solutions*

1. L. Fothergill to chief of the Special Projects Division, Chemical Warfare Service, "A Proposal for an American Plan for Postwar Research and Development of Biological Warfare," August 2, 1945, NA: RG 165, E. 488, B. 186.

2. Ibid.

3. Soon after the anthrax attacks, DOD officials began discussions with Merck to develop a next-generation vaccine. Merck indicated that, in contrast to BioPort, they had the capacity to produce sufficient quantities of the vaccine within months if Merck made the project a top priority. In the end, however, the DOD was unable to make an offer that was sufficiently attractive to encourage Merck to replace an existing product line with the anthrax vaccine or to build the new facilities required to manufacture it. Laura Johannes, Chris Adams, and Geeta Anand, "Health System on Alert: BioPort, Maker of Anthrax Vaccine, Struggles to Solve Regulatory Problems," *Wall Street Journal*, October 9, 2001, A. 6.

4. Irvin Stewart, *Organizing Scientific Research for War* (Boston: Little, Brown, 1948), 320.

5. While industry has been traditionally less motivated to pursue process innovations, there are some notable exceptions. Merck has recently developed a way to shorten development times and reduce failure rates by using biomarkers to establish clinical relevance of new drugs in humans earlier in the development process. Mervyn Turner, "Embracing Change: A Pharmaceutical Industry Guide to the 21st Century," in *Translational Medicine and Drug Discovery*, ed. Bruce Littman and Rajesh Krishna (Cambridge: Cambridge University Press, 2011).

6. Louis Galambos and Jane Eliot Sewell, *Networks of Innovation: Vaccine Development at Merck, Sharp & Dohme, and Mulford, 1895–1995* (Cambridge: Cambridge University Press, 1995), 250.

7. Andrei Shleifer, "State versus Private Ownership," *Journal of Economic Perspectives* 12, no. 4 (Fall 1998): 136.

8. World Bank, *Bureaucrats in Business* (London: Oxford University Press, 1995).

9. Aaron Friedberg, *In the Shadow of the Garrison State: America's Anti-Statism and Its Cold War Grand Strategy* (Princeton, NJ: Princeton University Press, 2000).

10. For a range of perspectives on firm boundaries, see Oliver E. Williamson and Sidney G. Winter, eds., *The Nature of the Firm: Origins, Evolution, and Development* (New York: Oxford University Press, 1993).

11. Peter S. Ring and Andrew H. Van de Ven, "Structuring Cooperative Relationships between Organizations," *Strategic Management Journal* 13 (1992): 438–498.

12. Ibid., 488.

13. Ibid., 491.

14. These professional networks included regulators as well. Joseph Smadel (former director of WRAIR's Department of Virus and Rickettsial Diseases, and mentor to Maurice Hilleman), for example, worked at the NIH from 1956 to 1963, eventually becoming chief of the Laboratory of Virology and Rickettsiology in the Division of Biological Standards.

15. Ring and Van de Ven, "Structuring Cooperative Relationships," 491.

16. Ibid., 490.

17. James Conant, president of Harvard University and chairman of the National Defense Research Committee, letter to the editor of the *New York Times*, August 3, 1945, in response to an unfavorable editorial regarding Vannevar Bush's proposals in Science: *The Endless Frontier*, copy in NA: RG 227, E. 2, B. 1.

18. Francis Collins, "Reengineering Translational Science: The Time Is Right," *Science Translational Medicine*, Volume 3, Issue 90, 90cm17, July 6, 2011.

19. The NIH's Vaccine Research Center (VRC) is an exception, although it is not designed to integrate process innovations.

20. Francis Collins, "Reengineering Translational Science: The Time Is Right," *Science Translational Medicine*, Volume 3, Issue 90, 90cm17, July 6, 2011, p. 2.

21. Research on creative problem solving shows that when individuals and teams are self-driven or "intrinsically motivated" they are best able to solve heuristic problems. Teresa Amabile, *Creativity in Context* (Boulder, CO: Westview, 1996); K. Lakhani and R. Wolf, "Why Hackers Do What They Do: Understanding Motivation and Effort in Free/Open Software Projects," in *Perspectives on Free and Open Software*, ed. J. Feller, B. Fitzgerald, and K. Lakhani (Cambridge, MA: MIT Press, 2005).

22. Defense Technical Information Center. DOD Dictionary of Military and Associated Terms. Joint Publication 1-02.

23. U.S. Marine Corps, Command and Control, MCDP 6, 104, www.dtic.mil/doctrine/jel/service_pubs/mcdp6.pdf

24. Peter Dombrowski and Eugene Gholtz, *Buying Military Transformation: Technological Innovation and the Defense Industry* (New York: Columbia University Press, 2006).

25. Ibid.

26. Ibid.

27. U.S. Department of Health and Human Services, "The Public Health Emergency Medical Countermeasures Enterprise Review: Transforming the Enterprise to Meet Long-Range National Needs," www.phe.gov/Preparedness/mcm/enterprisereview/Pages/default.aspx (accessed August 2010).

28. Anna Johnson-Winegar, "The DOD and the Development and Procurement of Vaccines against Dangerous Pathogens: A Role in the Military and Civilian Sector?" in Institute of Medicine, *Biological Threat and Terrorism: Assessing the Science and Response Capabilities: Workshop Summary* (Washington, DC: National Academies Press, 2002), 91.

29. As early as 1978, a government advisory group warned the Department of Health, Education, and Welfare that the high number of companies exiting the vaccine business was leading to the "dissolution of an expert staff and a removal of commitment to further activity in the field." "A Risky Exodus From Vaccines," *Business Week*, April 10, 1978.

30. ADMs could also leverage training and education programs to build private-sector surge capacity. In the same way that physicians commit to disaster medical assistance teams to respond to large-scale emergencies, ADMs could train and enlist advanced development and manufacturing personnel to participate in emergency manufacturing operations at federally funded ADMs.

31. See Appendix 3 for more details on military contributions to vaccine development.

Acknowledgments

This project began nearly fifteen years ago at MIT. Thanks to the early encouragement of Richard Danzig, the generous support of Paula Olsiewski at the Sloan Foundation, and the enthusiasm of Michael Fisher at Harvard University Press, it became a book. I am very grateful to all three. I am also indebted to the Program in the History and Social Study of Science and Technology at MIT, the Dibner Institute at MIT, and Harvard's International Security Program at the Belfer Center for Science and International Affairs. Each of these programs have supported and inspired research for this book at different junctures. I especially want to thank those brave souls who suffered through my early drafts: Richard Danzig, Catharina Wrede Braden, David Mindell, Harvey Sapolsky, Evelynn Hammonds, and Manning Rountree. David Franz, John Moon, Jeffrey Sturchio, Bruce Gellin, Greg Koblentz, Jeff Furman, Amy Salzhauer, Peter Wright, Andrew Price-Smith, Jennifer Leaning, Jenny Lind, Daryl Press, Cleo Sonneborn, and K. J. Dell'Antonia all provided perceptive comments on later drafts and did their best to set me straight. Phillip Russell deserves special mention for reading both early and late versions of this book. He endured this unenviable task with humor and provided valuable insights. The late Maurice Hilleman was equally entertaining, accommodating, enlightening, and altogether unforgettable. Don Metzgar, James Sorrentino, Tom Monath, Antonie Knoppers, Kenneth Bernard,

Patrick Kelley, Clem Lewin, Bill Egan, Charles Hoke, Joel Gaydos, Anne Kasmar, Chuck Ludlum, Peter Wright, Jess and Tim Lahey, Namita Sharma, and Lew Barker all generously shared their experiences, expertise, and insight. Finally, a huge thank you to my family for their abundant patience and support.

Index